Springer Tracts in Modern Physics
Volume 171

Managing Editor: G. Höhler, Karlsruhe

Editors: J. Kühn, Karlsruhe
Th. Müller, Karlsruhe
A. Ruckenstein, New Jersey
F. Steiner, Ulm
J. Trümper, Garching
P. Wölfle, Karlsruhe

Honorary Editor: E. A. Niekisch, Jülich

Now also Available Online

Starting with Volume 163, Springer Tracts in Modern Physics is part of the Springer LINK service. For all customers with standing orders for Springer Tracts in Modern Physics we offer the full text in electronic form via LINK free of charge. Please contact your librarian who can receive a password for free access to the full articles by registration at:

http://link.springer.de/series/stmp/reg_form.htm

If you do not have a standing order you can nevertheless browse through the table of contents of the volumes and the abstracts of each article at:

http://link.springer.de/series/stmp/

There you will also find more information about the series.

Springer
Berlin
Heidelberg
New York
Barcelona
Hong Kong
London
Milan
Paris
Singapore
Tokyo

http://www.springer.de/phys/

Springer Tracts in Modern Physics

Springer Tracts in Modern Physics provides comprehensive and critical reviews of topics of current interest in physics. The following fields are emphasized: elementary particle physics, solid-state physics, complex systems, and fundamental astrophysics.
Suitable reviews of other fields can also be accepted. The editors encourage prospective authors to correspond with them in advance of submitting an article. For reviews of topics belonging to the above mentioned fields, they should address the responsible editor, otherwise the managing editor.
See also http://www.springer.de/phys/books/stmp.html

Managing Editor

Gerhard Höhler

Institut für Theoretische Teilchenphysik
Universität Karlsruhe
Postfach 69 80
76128 Karlsruhe, Germany
Phone: +49 (7 21) 6 08 33 75
Fax: +49 (7 21) 37 07 26
Email: gerhard.hoehler@physik.uni-karlsruhe.de
http://www-ttp.physik.uni-karlsruhe.de/

Elementary Particle Physics, Editors

Johann H. Kühn

Institut für Theoretische Teilchenphysik
Universität Karlsruhe
Postfach 69 80
76128 Karlsruhe, Germany
Phone: +49 (7 21) 6 08 33 72
Fax: +49 (7 21) 37 07 26
Email: johann.kuehn@physik.uni-karlsruhe.de
http://www-ttp.physik.uni-karlsruhe.de/~jk

Thomas Müller

Institut für Experimentelle Kernphysik
Fakultät für Physik
Universität Karlsruhe
Postfach 69 80
76128 Karlsruhe, Germany
Phone: +49 (7 21) 6 08 35 24
Fax: +49 (7 21) 6 07 26 21
Email: thomas.muller@physik.uni-karlsruhe.de
http://www-ekp.physik.uni-karlsruhe.de

Fundamental Astrophysics, Editor

Joachim Trümper

Max-Planck-Institut für Extraterrestrische Physik
Postfach 16 03
85740 Garching, Germany
Phone: +49 (89) 32 99 35 59
Fax: +49 (89) 32 99 35 69
Email: jtrumper@mpe-garching.mpg.de
http://www.mpe-garching.mpg.de/index.html

Solid-State Physics, Editors

Andrei Ruckenstein
Editor for The Americas

Department of Physics and Astronomy
Rutgers, The State University of New Jersey
136 Frelinghuysen Road
Piscataway, NJ 08854-8019, USA
Phone: +1 (732) 445 43 29
Fax: +1 (732) 445-43 43
Email: andreir@physics.rutgers.edu
http://www.physics.rutgers.edu/people/pips/Ruckenstein.html

Peter Wölfle

Institut für Theorie der Kondensierten Materie
Universität Karlsruhe
Postfach 69 80
76128 Karlsruhe, Germany
Phone: +49 (7 21) 6 08 35 90
Fax: +49 (7 21) 69 81 50
Email: woelfle@tkm.physik.uni-karlsruhe.de
http://www-tkm.physik.uni-karlsruhe.de

Complex Systems, Editor

Frank Steiner

Abteilung Theoretische Physik
Universität Ulm
Albert-Einstein-Allee 11
89069 Ulm, Germany
Phone: +49 (7 31) 5 02 29 10
Fax: +49 (7 31) 5 02 29 24
Email: steiner@physik.uni-ulm.de
http://www.physik.uni-ulm.de/theo/theophys.html

Beate R. Lehndorff

High-T_c Superconductors for Magnet and Energy Technology

Fundamental Aspects

With 139 Figures

Springer

Dr. Beate R. Lehndorff
Universität Wuppertal
Fachbereich Physik
Gauss-Strasse 20,
42097 Wuppertal, GERMANY
E-mail: lehnb@uni-wuppertal.de

Library of Congress Cataloging-in-Publication Data applied for.

Die Deutsche Bibliothek - CIP-Einheitsaufnahme

Lehndorff, Beate R.:
High-T_c superconductors for magnet and energy technology: fundamental aspects/B. R. Lehndorff. –
Berlin; Heidelberg; New York; Barcelona; Hong Kong; London; Milan; Paris;Singapore; Tokyo: Springer, 2001
(Springer tracts in modern physics; Vol. 171)
(Physics and astronomy online library)
ISBN 3-540-41231-X

Physics and Astronomy Classification Scheme (PACS): 74.72.-h, 74.25.Fy, 74.62.Bf

ISSN print edition: 0081-3869
ISSN electronic edition: 1615-0430
ISBN 3-540-41231-X Springer-Verlag Berlin Heidelberg New York

This work is subject to copyright. All rights are reserved, whether the whole or part of the material is concerned, specifically the rights of translation, reprinting, reuse of illustrations, recitation, broadcasting, reproduction on microfilm or in any other way, and storage in data banks. Duplication of this publication or parts thereof is permitted only under the provisions of the German Copyright Law of September 9, 1965, in its current version, and permission for use must always be obtained from Springer-Verlag. Violations are liable for prosecution under the German Copyright Law.

Springer-Verlag Berlin Heidelberg New York
a member of BertelsmannSpringer Science+Business Media GmbH

© Springer-Verlag Berlin Heidelberg 2001
Printed in Germany

The use of general descriptive names, registered names, trademarks, etc. in this publication does not imply, even in the absence of a specific statement, that such names are exempt from the relevant protective laws and regulations and therefore free for general use.

Typesetting: Camera-ready copy by the author using a Springer LATEXmacro package
Cover design: *design & production* GmbH, Heidelberg

Printed on acid-free paper   SPIN: 10765115   56/3141/tr   5 4 3 2 1 0

To my family

L'essentiel est invisible pour les yeux.
(What is essential is invisible to the eye.)
Antoine de Saint-Exupéry: *Le Petit Prince*
(translated by Katherine Woods)

Preface

Since the discovery of high-temperature superconductors the scientific community has been very active in research on material and system development as well as on the basic understanding of the mechanism of superconductivity at high transition temperatures. Industrial groups joined in very soon as with these new materials the prospects for commercial application of superconductivity seemed to be more promising than ever. Materials processing was divided into film deposition and bulk preparation techniques, the latter including conductor fabrication and melt growth of monolithic samples as well. Because of the high impact of possible applications in energy technology, wire and tape fabrication of the BSCCO superconductors is one of the most important fields, in addition to thin-film technology for mobile comunication. Only since processes like IBAD and RABiTS^{TM} were invented have film deposition techniques also become important for energy technology.

In order to produce suitable conductors with material properties which meet the challenge imposed by energy technology, detailed understanding of the phase formation and physical properties of the high-temperature superconductors is necessary. The goal of this book is on one hand to provide the basic information on phase formation and physical properties, and to give a short overview of the state of the art in conductor preparation and characterization. On the other hand it contains the author's own results in the field of preparation and characterization. This work has been performed at the Bergische Universität Wuppertal and at the Applied Superconductivity Center of the University of Wisconsin at Madison. It has been partly funded by the BMBF and the Ministry of Science of North Rhine–Westphalia. I want to acknowledge the support and collaboration of many people. I am grateful to Prof. Helmut Piel for the possibility to work in this exciting field in his laboratory. I appreciated very much the collaboration with Hans-Gerd Kürschner, Dr. Michael Hortig, Jahangir Pouryamout, Bernhard Lücke, Dr. Bernhard Fischer, Rainer Wilberg, Bernd Günther, Peter Hardenbicker, Markus Getta, Sascha Kreiskott, Stefan Hensen and Prof. Günter Müller and all my other colleagues at the University of Wuppertal. I am also indebted to Prof. David Larbalestier for the possibility to work in his group at the Applied Superconductivity Center of the University of Wisconsin at Madison. During that year I enjoyed very much the inspiring atmosphere in his group. In addition

I want to thank all colleagues in Madison, especially Jodie Reeves, Dr. Wei Zhang, Prof. Eric Hellstrom, Dr. Mark Rikel and Dr. Anatolii Polyanskii, for their collaboration during my stay and afterwards. Last but not least I want to thank my family for their patience.

Wuppertal, August 2000 *Beate Lehndorff*

Contents

1. Introduction 1
2. High-T_c Superconductors: Limitations and Applications .. 5
   2.1 Critical-Current Limitations 6
      2.1.1 Flux Pinning 7
      2.1.2 The Role of Grain Boundaries 8
      2.1.3 Percolation 10
   2.2 Material Processing 11
   2.3 Possible Applications 11
      2.3.1 Magnet Technology 11
      2.3.2 Energy Technology 12

Part I. Fundamentals

3. Fundamentals of Material Processing 17
   3.1 Crystal Structure 18
      3.1.1 The Rare-Earth Compounds 18
      3.1.2 The Bismuth Compounds 19
      3.1.3 Lattice Parameters, Anisotropy and Microstructure ... 20
   3.2 Phase Diagrams and Phase Formation 22
      3.2.1 Basic Concepts 22
      3.2.2 Binary Phase Diagrams 24
      3.2.3 Ternary Phase Diagrams 26
      3.2.4 The Phase Diagram of Y–Ba–Cu–O 28
      3.2.5 The Phase Diagram of Bi(Pb)–Sr–Ca–Cu–O 30
      3.2.6 Oxygen Content 33
   3.3 Melt Processing of $YBa_2Cu_3O_{7-\delta}$ 34
      3.3.1 Melt Growth Processes 35
      3.3.2 Top-Seeding Technique 38
   3.4 Tape and Wire Fabrication 38
      3.4.1 Conductor Fabrication Processes 38
      3.4.2 Green-Wire Fabrication Using the PIT Process 39
      3.4.3 Phase Development of Pb,Bi–Sr–Ca–Cu–O in Silver .. 42
      3.4.4 YBaCuO-Coated Conductors 46

4. Physical Properties of High-T_c Superconductors ... 51
4.1 Normal-State Properties of HTSCs ... 51
4.2 Superconducting Properties ... 52
4.2.1 Microscopic Description ... 52
4.2.2 Macroscopic Description ... 54
4.3 Electromagnetic Properties of HTSCs ... 57
4.3.1 Type II Superconductor in an External Magnetic Field ... 57
4.3.2 Elastic Properties of the Flux Line Lattice ... 59
4.3.3 Phase Transitions in the Vortex Lattice ... 60
4.4 Flux Line Dynamics ... 62
4.4.1 Flux Pinning and the Bean Critical-State Model ... 62
4.4.2 Thermally Activated Flux Creep ... 65

Part II. Recent Achievements

5. Conductor Preparation and Phase Evolution ... 71
5.1 Preparation of Wires, Tapes and Bulk HTSCs ... 71
5.1.1 Preparation of Bi-2223 Tapes ... 71
5.1.2 Preparation of Bi-2212 Tapes and Wires ... 74
5.1.3 YBCO-Coated-Conductor Preparation ... 75
5.1.4 Melt Processing of Bulk REBCO ... 76
5.2 Phase Formation and Microstructure ... 76
5.2.1 Observation of Phase Evolution of Bi-2223 ... 76
5.2.2 Microstructure and Grain Boundaries ... 77

6. Characterization of Conductors and Bulk HTSCs ... 79
6.1 Electromagnetic Characteristics ... 79
6.2 Superconducting Magnetic Levitation ... 80
6.3 Imaging of Magnetic Flux in Type II Superconductors ... 80
6.3.1 Scanning Hall Probe Experiments ... 81
6.3.2 Magneto-Optical Imaging ... 81

Part III. Phase Formation and Microstructure

7. Preparation of BSCCO Conductors ... 85
7.1 Fabrication of Green Wires and Tapes ... 85
7.2 Thermal Processing of Bi-2212/Ag Conductors ... 88
7.2.1 Processing Scheme ... 89
7.2.2 Void Formation ... 89
7.2.3 Microstructure Development ... 92
7.3 Thermo-Mechanical Processing of Bi-2223/Ag Tapes ... 93
7.3.1 Precursor Powder ... 94
7.3.2 Number of Processing Steps ... 95

		7.3.3 First Thermal Processing Step 96

   7.3.3 First Thermal Processing Step 96
   7.3.4 Further Processing Steps 99

8. Overpressure Processing of Bi-2212 Conductors 101
   8.1 Void Reduction in Bi-2212/Ag Wires...................... 101
      8.1.1 Processing Scheme................................ 101
      8.1.2 Critical Current Density.......................... 102
      8.1.3 Microstructure 102
   8.2 Overpressure Processing of Bi-2212/Ag Tapes 103
      8.2.1 Thickness Dependence of the Critical Current Density 104
      8.2.2 Magneto-Optical Imaging 105
      8.2.3 Microstructural Analysis 105
      8.2.4 X-ray Results 107
      8.2.5 Interpretation.................................... 108

**9. Processing of Bi-2223/Ag Tapes
   at Reduced Final Temperature** 109
   9.1 Processing Schemes 109
   9.2 Critical Current Density................................. 110
   9.3 X-ray Analysis 110
   9.4 Microstructural Examination 112
   9.5 Ac Susceptibility Results 112
   9.6 Discussion 114

10. Preliminary Results on YBCO-Coated Conductors 117
   10.1 Metallurgy of the Metallic Substrates 117
      10.1.1 Recrystallization Procedure........................ 117
      10.1.2 Texture Analysis 120
   10.2 Buffer Layers 122
      10.2.1 Buffer Layers on Nickel Substrates.................. 122
      10.2.2 Buffer Layers on Ni–Cu Composite Tapes 124
   10.3 Y–Ba–Cu–O Coating 124

Part IV. Electromagnetic Properties

11. Magnetotransport and Vortex Dynamics.................. 129
   11.1 Magnetic-Field Dependence of the Critical Current Density .. 129
   11.2 Scaling of the I–V Characteristics 130
   11.3 Critical-Current Anisotropy.............................. 133
   11.4 Flux Creep Resistivity 135
   11.5 Irreversibility Field 136

12. Superconducting Magnetic Levitation 139
   12.1 Levitation Force Experiments on Various Superconductors ... 139
      12.1.1 Experimental Setup 139
      12.1.2 Experimental Parameters........................... 140
      12.1.3 Comparison of Various Superconductors 146
      12.1.4 Spatial Distribution of Levitation Force 149
   12.2 Understanding Force–Distance Hysteresis 151
      12.2.1 Interpretation Within the Bean Critical-State Model .. 152
      12.2.2 Magnetic-Field-Dependent Critical Current Density ... 156
   12.3 Interpretation of Vertical-Stiffness Experiments............. 157
      12.3.1 Vertical Magnetic Stiffness as a Bearing Parameter ... 158
      12.3.2 Stiffness and Labusch Parameter 159
   12.4 Experiments on Stacks of Epitaxial Thin Films............. 163
      12.4.1 Force–Distance Characteristics 163
      12.4.2 Magnetic Stiffness 164

13. Remanent Flux Distribution and Critical Current Density 167
   13.1 Experimental Details 167
      13.1.1 The Hall Sensor 167
      13.1.2 Resolution of the Hall Probe 168
      13.1.3 Imaging of the Remanent Flux Distribution.......... 169
   13.2 Calculation of the Critical Current Density 169
   13.3 Epitaxial YBCO Films 171
      13.3.1 Experimental Details 171
      13.3.2 Results and Discussion 172
   13.4 Bi-2223/Ag Tapes 172
      13.4.1 Lateral Resolution.................................. 173
      13.4.2 Results from Bi-2223/Ag Tapes 175
      13.4.3 Quality Control of Long Tapes 178

Part V. Concluding Remarks

14. Concluding Remarks 181
   14.1 Summary... 181
   14.2 Conclusion 182

References 187

Index 203

1. Introduction

When Heike Kammerlingh Onnes found superconductivity in pure mercury in 1911 [118] after liquefying helium in 1908 [117] he immediately knew about the great potential of lossless current transport for applications in magnet and energy technology. The second important property of a superconductor besides carrying current without resistance is the *Meissner–Ochsenfeld effect* [192], i.e. magnetic flux is completely expelled from the interior of a metal when it goes to the superconducting state. The first decades of superconductor research were spent in finding theoretical models for this new phenomenon and on the search for materials with higher transition temperatures. The brothers Fritz and Heinz London proposed an electromagnetic theory in 1935 [174] explaining the finite penetration length of electromagnetic fields into the superconductor (*London penetration depth*). In 1950 Ginzburg and Landau formulated a theory based on the thermodynamics of phase transitions [67] which gives a good macroscopic description of the superconducting state. A detailed microscopic theory was given by Bardeen, Cooper and Schrieffer in 1957 [12]. This *BCS theory* assumed electron–phonon coupling as the underlying mechanism for the pairing of conduction electrons into *Cooper pairs*, which carry the supercurrent.

Many pure metals were found to be superconducting, with Nb having the highest transition temperature for a pure element of 9.2 K. But on the route to application these metals had a great disadvantage. At a fairly low external magnetic field (*the critical field H_c*) of some ten mT superconductivity is destroyed because the magnetic energy exceeds the coupling energy of the Cooper pairs. The first way out of this dilemma was found with the A15 compounds. A well-known material from this class is Nb_3Sn, which was found by Bernd Matthias in 1954 [189]. The A15 compounds have the complicated β-tungsten structure and show transition temperatures well above 10 K. For example Nb_3Sn has $T_c = 18$ K, Nb_3Ge has the highest T_c of these compounds, with 23 K, and V_3Ga can reach a T_c of 20 K when prepared carefully.

These new materials introduced a new type of superconductivity, thus leading to the discrimination between *type I* and *type II* superconductors. In a type I superconductor superconductivity is stable up to the critical filed H_c. A type II superconductor has a lower and an upper critical field H_{c1} and H_{c2}. When the lower critical field is reached it becomes energetically favorable to let magnetic flux penetrate the superconductor. This happens in the form of quantized vortices called *Abrikosov flux lines*. The transition to the normal-conducting state happens only at H_{c2}, which is nearly 20 T for Nb_3Sn, i.e. several orders of magnitude higher than in pure metals. Thus a type II superconductor can occupy two distinct superconducting states, the *Meissner state*, where no flux is present in the interior of the material, and a mixed phase, also called *the Shubnikov phase*, where quantized vortices penetrate the superconductor.

Abrikosov [2] could show, in 1957, that the Shubnikov phase is a periodic solution of the Ginzburg–Landau (GL) equation and the energetically most probable symmetry is triangular. The experimental verification of the triangular Abrikosov *flux line lattice* (FLL) was performed by Essmann and Träuble in 1967 [49] using a magnetic decoration technique. In addition Gorkov could prove in 1960 [73] that the GL theory is an exact solution of the BCS theory near to the superconducting phase transition, extending this theory to the so-called GLAG theory.

In order to take advantage of the high upper critical fields of the A15 compounds for the design of high-field magnets an additional problem had to be solved. When a current is flowing in a type II superconductor vortices can move freely, thus dissipating energy, unless they are pinned to a fixed site in the crystal lattice. A superconductor with good flux-pinning properties is called *hard* superconductor. Only these special materials are suitable for applications, as only they are able to carry the high currents necessary to produce magnetic fields of a strength of several tesla.

Long, flexible wires are needed to build a high-field magnet. However, Nb_3Sn is a brittle material and it is not an easy task to fabricate wires out of it. This technical problem could not be solved until the alloy NbTi was found. Despite its lower transition temperature of 10.5 K and its smaller upper critical field of about 12 T, it offers the significant advantage of flexibility and formability. Wires of NbTi in a copper matrix can easily be produced in long lengths and wound into magnets. Magnets wound from NbTi wires became the first commercially available applications of superconductivity and came to market in the 1960s, fifty years after Kammerlingh Onnes first reported superconductivity in mercury. Nowadays these magnets are widely used in magnetic resonance imaging (MRI) and offer a profitable market, mainly in health care but also in research. These magnets have to be cooled with liquid helium, which is not a problem in stand-alone systems. However, for applications in energy technology like power transmission cables, cooling with liquid helium turned out to be uneconomic. Thus magnet technology still is the only market for superconductivity at present.

Matthias gave a very useful rule for the transition temperature of the conventional superconductors [188], which tells us that T_c increases with increasing average number of valence electrons of a metal. This led to the assumption that transition temperatures above 30 K were impossible to achieve

1. Introduction

in any material. This opinion had to be changed in 1986 when Bednorz and Müller found superconductivity at 40 T in LaSrCuO [15]. K.-A. Müller was already well known from his work on ferroelectric perovskites. Inspired by an idea of Bernd Matthias, who suspected that owing to the enhanced electron–phonon coupling in ferroelectric perovskites, this class of material might offer a way out of the limitation on T_c [202], he and Bednorz synthesized different compounds, finally finding LaSrCuO. This work caused a boom in research on superconductivity, and a flood of papers was published during the following years. The highlights were the detection of YBaCuO with $T_c = 93$ K by Chu's group [272] half a year after Bednorz and Müller's discovery, and the Bi compounds BiSrCaCuO with T_cs of 86 K and 110 K by Maeda et al. in 1988 [179]. The new materials have very high upper critical fields of more than 50 T; however, they have turned out to be even more brittle than Nb_3Sn as they are ceramics. Consequently the initial hope of realizing additional commercial applications, especially in energy technology, and to earn big money from this new market had soon to be scaled down. The big problem is: how can one form a flexible wire out of a ceramic? Thus, after ten years of high-T_c superconductivity, the new materials still are no serious challenge to the conventional superconductors. The goal of this work is to throw some light on the reasons for these difficulties.

Many new theories different from BCS emerged in order to explain the high transition temperatures of the oxide superconductors. It cannot be the aim of this work to go into great detail about these theories. Furthermore, the new superconductors are compounds of four or more constituents, leading to very complicated phase diagrams. Consequently phase formation in these materials is complex. As this field of research is fairly new, much of the knowledge necessary to start working in this field is found not in textbooks but in original papers. For this purpose, in the first part of this book some basic facts about phase formation and electromagnetic properties of high-T_c superconductors are put together. This part should serve as a short overview for newcomers in this field, e.g. students starting with their diploma or PhD thesis work. In the second and third parts, my own contributions to this very competitive field are presented. In these parts the results presented are meant to show the problems materials scientists are still facing when preparing ceramic superconductors for magnet and energy technology. The fundamental and results parts are preceded by a chapter in which the main challenges imposed by the new superconductors are addressed and some possible applications are presented.

2. High-T_c Superconductors: Limitations and Applications

When high-temperature superconductivity was found by Bednorz and Müller in 1986 [15] many research activities were started all over the world. This was not only because of the desire for even higher critical temperatures T_c. But the possibility of cooling with liquid nitrogen instead of liquid helium made the application of superconductivity in energy technology much more probable and economically feasible. The great expectations, however, had soon to be scaled down. Despite having high transition temperatures of e.g. 94 K in $YBa_2Cu_3O_{7-\delta}$ or even 110 K in $(Pb,Bi)_2Sr_2Ca_2Cu_3O_{10}$, the new ceramic superconductors could not carry very high critical currents I_c. Thus technical applications again turned out not to be around the corner and big effort was put into the understanding and elimination of critical-current limitations.

Yet there are some important properties of high-temperature superconductors (HTSCs) besides the high transition temperature. In principle they have a very high upper critical field (up to 100 T for the bismuth superconductors) and could thus carry high currents at much higher fields than the conventional metallic superconductors. In addition to application in energy technology this would offer the possibility of building very-high-field magnets which would generate fields of over 20 T in the laboratory, the limit imposed by conventional superconductors at present.

To get an impression of how the current-carrying capacities of the various materials compare in a magnetic field and at different temperatures it is useful to draw all j_c versus B dependences in one figure. This is done in Fig. 2.1. From this presentation it can be seen that the critical current density at 4.2 K and in elevated magnetic fields of $Bi_2Sr_2CaCu_2O_8$ (Bi-2212) tapes outreaches that of NbTi and even Nb_3Sn. But it can also be seen that the magnetic-field dependence of j_c for $(Pb,Bi)_2Sr_2Ca_2Cu_3O_{10}$ (Bi-2223) tapes at 77 K is very strong. Even at about 1 T j_c has dropped below 1 kA/cm^2, with severe consequences for the application of this material, namely that high magnetic fields cannot be generated at 77 K with Bi-2223. This, however, already shows the reasons for the great potential impact of finding out the origin of the low current-carrying capacity of the new materials.

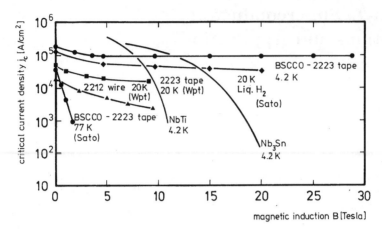

Fig. 2.1. Critical current density as a function of the external magnetic field for different superconductors

2.1 Critical-Current Limitations

Why is the current-carrying capacity so limited in the new superconductors? The HTSCs are ceramics and are thus brittle materials consisting of many more or less well-connected grains. The supercurrent has to flow within these grains and over the grain boundaries, which necessarily leads to certain limitations in the current-carrying capacity.

There are several physical reasons for this limitation on the critical current, or on the critical current density j_c, which is normally used in the literature as it is independent of the conductor cross section. The main sources have been given by Larbalestier [153] and are summarized in Table 2.1.

Table 2.1. Limiting critical current densities in polycrystalline high-T_c superconductors

Origin	Magnitude
Pair breaking	$j_d = 10^8 - 10^7$ A/cm^2
Flux pinning	$j_{fp} = 10^7 - 10^6$ A/cm^2
Grain boundary	$j_g = 10^7 - 10^2$ A/cm^2
Percolation	$j_p = 10^5 - 10^4$ A/cm^2

The pair-breaking critical current density is far from being reached in most polycrystalline materials. Only the last three sources are really relevant in technical superconductors, and will be discussed in some detail. In the case of alternating currents the high ac losses have to be considered in addition.

2.1.1 Flux Pinning

The intragrain critical current density seems to be at least one order of magnitude higher than the intergrain j_c. This can be concluded from experiments on the ground core of Bi-2223/Ag tapes [32]. The limitation on j_c within the grains is caused by the motion of magnetic flux lines in the Shubnikov phase of the superconductor. In an ideal type II superconductor the critical current density is infinitely small because flux lines can move freely. In a real superconductor there are always lattice defects and/or impurities which serve as pinning centers for the flux lines. The better the flux-pinning properties the higher the critical current density in a so-called hard superconductor. The limit j_c is reached when vortices can leave their pinning sites and move through the lattice, dissipating energy. This subject will be discussed in some detail in Sect. 4.3. The material development of the conventional low-temperature superconductors (LTSCs) like NbTi or Nb_3Sn for applications in magnet technology was based on the engineering of pinning properties. The structure of vortices in these metallic materials is quite simple compared with those in the HTSCs. Nevertheless the development of flexible wires with appropriate pinning properties took about 50 years to reach the first commercially available magnets.

The structure of flux lines in the anisotropic ceramic superconductors is much more complicated. While in LTSCs a vortex is really a flux line, which can be pinned by just one or a few pinning sites, the vortices in HTSCs can split into two-dimensional pancake vortices. They are very weakly coupled and have to be pinned separately. This means that the pinning-site density has to be rather high but without destroying the superconducting properties.

Further difficulties arise from the fact that the superconducting coherence length ξ in the ceramic superconductors is quite small (0.3 to 3 nm) as compared with conventional materials like Nb (40 nm) or Pb (500 nm). Pinning centers act most efficiently if their spatial extension is of the order of the coherence length. This implies that in the HTSC materials pinning sites of atomic size have to be introduced. The first results were achieved by radioactive irradiation, e.g. of Bi-2223 tapes [144, 145]. Recent results of the group at Brookhaven National Laboratory, however, show that there is a limitation to this method [25, 167]. In a magnetic field above 4 T the pinning centers created by irradiation damage become ineffective. Furthermore, irradiation doses of more than 4 T "dose-equivalent field" are deleterious and cause a significant decrease of the superconducting transition temperature. Although of some advantage, this method has the great disadvantages that it is too expensive and the remaining radioactive contamination is considered to be hazardous.

A different route, widely used in conventional superconductors, is the creation of pinning centers via chemical precipitation. Because of the complicated phase diagrams of HTSCs, which will be discussed in more detail in Sect. 3.2, this is not an easy way, though. For example, in the case of $YBa_2Cu_3O_{7-\delta}$ small inclusions of the nonsuperconducting compound Y_2BaCuO_5 serve as efficient pinning centers. Most likely, the lattice deformations in the neighborhood of the inclusions have a suitable size. The first positive results were reported on lead doping of Bi-2212 single crystals [249]. There was also an attempt to transfer these results to Bi-2212 silver-clad tapes but the critical current densities obtained to date are not very high [61]. More convincing results have been reported by Anderson, who varied the lead content of Bi-2223/Ag tapes, which led to a change in the irreversibility field [7].

There is a discussion in the literature as to whether intergrain or intragrain currents limit the current-carrying capability of the ceramic superconductors. As it is possible to improve the critical current density by irradiation damage, Hensel et al. conclude [95] that in high-j_c tapes of an HTSC the grain boundaries are no longer weak links and the critical current is limited only by intrinsic mechanisms, i.e. flux pinning. The magnetically determined j_c of ground cores of HTSC tapes, however, is one order of magnitude higher than the transport critical current density of the whole tape, as Caplin et al. could show [32, 40]. This is an indication that weak links limit the current-carrying capacity.

Nevertheless there is also a physical constraint in principle on the flux-pinning properties of these materials. As the transition temperatures are rather high, around 100 K and more, thermally activated flux creep is thermodynamically more important.

2.1.2 The Role of Grain Boundaries

The above-mentioned discussion about inter- and intragrain contributions to the critical current density reveals the important role of the grain boundaries in these materials. Owing to the very short coherence length of 0.3 to 3 nm in the ceramic superconductors, only small barriers between grains are necessary to change a grain boundary from a strong link to a weak link. These barriers may be impurities, lattice distortions or oxygen deficiencies. Also, misorientation between adjacent grains can lead to weak-link behavior. In YBCO there exists a crossover misorientation angle for this behavior, as could be shown by Heinig et al. [91]. As a consequence it appears to be necessary to obtain a material with only low-angle grain boundaries, as they most likely act as strong links.

Weakly coupled grains decouple in increasing external magnetic fields. This leads to an initial drop of the critical current density at fields below 1 T, which is readily observed in most HTSC materials above 30 to 40 K. Obviously this is a great disadvantage for some applications, like magnet design.

Several models have been developed to explain current transport in these multiply connected materials. These models have to make certain assumptions as to the nature of the grain boundaries crossed by the current flow.

The first transport model – the so called *brick wall model* – was proposed by Malozemoff in 1990 [184] and extended by Bulaevskii and coworkers in 1993 [26]. In this model it is assumed that the grains are arranged like bricks in a wall, as shown on the left-hand side of Fig. 2.2. This assumption stems from the fact that the real microstructure of a polycrystalline HTSC consists of thin platelets extending several hundred microns along the crystal ab planes but only a few microns thick in the c direction. Thus c axis contacts have much larger contact areas, and current transport through these c axis boundaries should be the only relevant current transport. One argument against this model is that, from the results of Kleiner and Müller [129], electrical transport along the c axis is intrinsically weakly coupled.

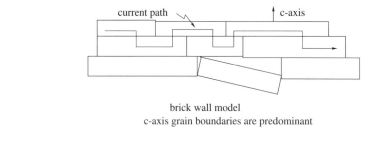

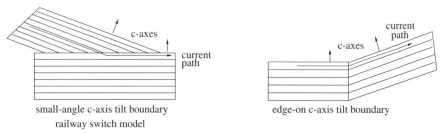

Fig. 2.2. The most common grain arrangement in BSCCO tapes, as assumed by the brick wall model (*left*) and the railway switch model (*right*)

In 1995 Hensel et al. proposed the *railway switch model* as an alternative view of current transport over grain boundaries [95]. The basic assumptions here are that the grains are always slightly misoriented and electrical transport can only take place via edge-on ab contacts. That means that one misoriented grain meets another like a railway switch and current can flow from one ab plane to the next. This model is depicted on the right-hand side of Fig. 2.2. In this model c axis transport is not considered to be important. Hensel argues that in high-quality tapes only low-angle grain boundaries exist. They are all strongly coupled and thus flux-pinning limits the critical current density. There can also be found results that contradict this model. In their experiments on extracted single filaments of Bi-2223/Ag multifilamentary tapes Cai et al. [27] showed that there can still be found signatures of c axis transport in the I–V curves.

A combination of both models is represented by the *freeway model* proposed by Riley et al. in 1996 [233]. Within this model the current can find its way into and out of a grain preferentially through the contact which has the lowest resistance. If there is no ab contact with a neighboring grain a c axis contact has to serve as the current path.

It is most likely that a more general view of current transport across grain boundaries has to be taken into consideration. If one takes a closer look at the real microstructure of a ceramic superconductor in an electron microscope all sorts of grain boundaries can be found. Still, this microstructure has to be connected to results about current transport across certain grain boundaries. This connection between critical current density and microstructure remains a great challenge and plays a crucial role in the understanding of current transport in HTSCs.

2.1.3 Percolation

Flux pinning and grain boundaries have a strong influence on the critical current densities in high-temperature superconductors. But, comparing the values of j_c of Table 2.1 with the currently reported values of $70\,\text{kA/cm}^2$ [234] in short samples of Bi-2223/Ag tapes or $25\,\text{kA/cm}^2$ [83] in longer tapes, it appears that these experimental values have as their origin j_p, the percolation critical current density. Looking at the real microstructure of the core of a Bi-2223/Ag tape, one sees that it does not only consist of grains and grain boundaries, but there are also spurious second phases and even microcracks and voids. Considering the short coherence length, second phases and voids of a few hundred nanometers in size are big enough to totally block current transport between grains, leading to a percolative current path. This has been nicely demonstrated by magneto-optical investigations [216]. Even in the very-high-j_c tapes there are microcracks blocking the current path [27]. As a consequence only parts of the total cross section of the tape carry current. It is assumed that the active current-carrying area is as low as 2% of the total cross section used for calculating the critical current density from the measured quantity I_c.

As can be seen from the chemical formula, HTSC materials are compounds of four to six elements. This leads to very complex phase diagrams, as will be presented in detail in Sect. 3.2. Second-phase precipitation during processing is therefore hard to avoid. There are also different reasons for the origin of microcracks, voids and porosity, like the mechanical treatment between stages of thermal processing, emanation of gases during heat treatment or volume reduction due to melt processing. Some of the mechanisms could be identified by Zhang et al. [284] for the processing of Bi-2212 tapes. These topics will be discussed in more detail in Chap. 8.

2.2 Material Processing

As already mentioned the high-T_c superconductors are compounds of four – like $YBa_2Cu_3O_{7-\delta}$ – or even six constituents – like $(Pb,Bi)_2Sr_2Ca_2Cu_3O_{10}$. As a consequence the phase diagrams are very complicated and can be understood only in part. Phase formation in such a multicomponent system is accompanied by many coexisting phases, leading to second-phase precipitates. The situation becomes even more complex when it comes to conductor formation. The HTSCs are ceramics and thus brittle materials. Therefore wires cannot easily be formed as in the case of NbTi, for example. A ductile matrix material had to be found. It turned out that the only metal which could be used for this purpose was silver, as it does not react with the superconducting ceramic. In addition its high permeability to oxygen enables the necessary exchange between the reacting core and ambient atmosphere. However, the silver cladding adds a further component to the system which has to be taken into account.

HTSC tape fabrication needs various basic processes which have to be controlled carefully. Mechanical deformation, in the case of Ag/BSCCO composites, is necessary to form the green tapes which subsequently undergo thermal processing for phase formation. In the case of YBCO-coated conductors various layer deposition techniques are applied. These processes are as important as the phase formation and have to be controlled carefully. Owing to the importance of the materials processing, Chap. 3 will treat fundamentals like phase diagrams and conductor formation in detail.

2.3 Possible Applications

A great source of motivation for the material development is commercialization in useful systems built out of HTSCs. In principle there are two major fields of application for bulk and conductor material made from ceramic superconductors: magnet technology and energy technology.

2.3.1 Magnet Technology

Magnet technology is the traditional field of superconductivity and the first commercial applications of conventional superconductors were in the field of magnets for laboratories and particle accelerators. With the HTSCs a further possible application is magnets which can be operated without a cryogenic liquid, using a closed-cycle refrigerator. These magnets will run at temperatures around 30 K, which is the ideal region for refrigerators. But the development of refrigerators has been very busy since the arrival of HTSC ceramics. They now reach even lower temperatures with reasonable cooling power. Thus the first cryogen-free magnet was built out of NbTi with HTSC current leads [112], running at 4.2 K.

Still, the competition between conventional and high-T_c superconductors is very hard. The only opportunity for HTSC magnets will be in fields where the conventional superconductors definitely are at their limit. These areas of application are very-high-field magnets and particle accelerators. The bismuth superconductors are able to reach much higher magnetic fields than, e.g., Nb_3Sn without becoming normal-conducting. Thus much effort is being put into the development of hybrid magnets consisting of NbTi and Nb_3Sn magnets with insert coils built from $Bi_2Sr_2CaCu_2O_8$. Such a high-field magnet is interesting for high-resolution NMR experiments and has been developed by Hitachi in Japan in collaboration with the NRIM (National Research Institute of Metals, Japan) [143].

Particle accelerators need magnets for beam deflection (dipole magnets) and focusing (quadrupole magnets). The higher the magnetic fields which can be reached with the dipole magnets, the smaller a particle accelerator can be built to obtain the same particle energy or the higher the particle energy that can be reached by an accelerator of given size. While very-high-field magnets are simple solenoids, dipole magnets have a more complicated shape. The magnetic field strength produced by dipole magnets is of the order of 7 T. A magnetic field of this strength causes a pressure of more than one hundred atmospheres. As a consequence mechanical stabilization is a major problem in magnet design for accelerators. Conductors for this application need to have very high mechanical stability.

At present the only application of HTSCs which could lead to commercialization in the near future is current leads, especially for accelerator magnets. These current leads are normally built from copper and consequently introduce much heat into the system, which has to be cooled. HTSC current leads would reduce the heat input by an appreciable amount. The Large Hadron Collider (LHC), which will be built in Geneva, will have dipole and quadrupole magnets using conventional superconductors. The current leads are planned to be fabricated from HTSCs [250].

2.3.2 Energy Technology

The idea of lossless current transport in superconductors has encouraged much development work in the field of energy technology. Energy transmission cables made from HTSCs could meet the increasing energy demand of big cities like Tokyo using the old underground lines of the conventional copper cables. It was estimated that three times the electric power could be provided within the same underground cable tunnels by replacing the copper transmission cables by HTSC conductors. For that reason power transmission cables built from ceramic superconductors are being developed in Japan by, among others, the Tokyo Electric Power Company (TEPCO) with great effort [238, 248].

Other ac systems in the power grid suitable for operation with superconductors are transformers, motors and generators. The efficiency of those

2.3 Possible Applications

systems could be enhanced by a few percent, which would provide sufficient power savings over the long lifetime of, e.g, a transformer. But crucial for the operation of cables and transformers are safety systems which prevent damage in the case of a fault current. At present, fault current limiters are explosive switches which disconnect a line carrying a fault current from the grid. These current limiters are nonrecoverable systems and have to be replaced after a fault current event. A superconducting fault current limiter can be a recoverable device as it utilizes the superconducting transition to limit the fault current. Such systems are being developed in several parts of the world. One prototype, as well as a prototype transformer built of HTSCs, has been operated in a power grid for one year in Geneva by ABB [217]. However, the high ac losses of the ceramic superconductor still hinder their commercial application in this field.

Magnetic levitation (MAGLEV) offers a further possibility for the application of superconductivity. Magnetically levitated trains are being built in Germany (TRANSRAPID [191]) and Japan [259]. But only the Japanese system uses superconducting magnets to achieve levitation. At present only conventional superconductors are being considered for these magnets. HTSCs are being considered as screening devices for the passenger cabins.

Last but not least, it should be mentioned that superconductivity can also be used in energy storage devices. There are two possibilities to store energy using a superconductor: energy storage in a magnetic field (SMES: superconducting magnetic energy storage) and storage as mechanical energy in a flywheel. Both systems are being developed. While an SMES under development [116] uses conventional as well as high-T_c superconductors, flywheels [19] are being built as rotational magnetic bearings using bulk $YBa_2Cu_3O_{7-\delta}$. Their field of application is mainly as uninterrupted power supplies (UPS) for sensitive companies or power grids. There is not enough space in this book to discuss all the possible applications of superconductivity in magnet and energy technology in detail, as the book is focused on fundamental aspects. Two books, by T. Sheahen [247] and P. Komarek [135], which provide more insight are therefore recommended to the interested reader.

All these systems have a few things in common. For their fabrication, a long length of conductor or a large amount of good-quality bulk material is needed. Fabrication processes have to be found which provide reliable standard-quality material with reproducible mechanical and superconducting properties, sufficiently high critical current densities and affordable costs. This is not an easy requirement to meet. On the way a lot of work has to be done in understanding the specific properties of the material, and preparation processes have to be developed. It is the purpose of this book to provide some of the basic information necessary for understanding and optimizing phase formation in HTSCs as well as their special electromagnetic properties.

Part I

Fundamentals

3. Fundamentals of Material Processing

The high-T_c superconductors are complex compounds with four to six constituents. Their crystal structure is anisotropic, leading also to anisotropic physical properties. In order to optimize the materials processing and superconductivity these properties need to be well understood. In detail, a basic knowledge of the crystal structure, of the thermodynamics of the phase diagrams and phase formation, and of the normal and superconducting properties is necessary. In particular, the thermodynamics and kinetics of phase formation in Bi–Sr–Ca–Cu–O silver-clad tapes need to be thoroughly looked at, but it has to be emphasized that not all mechanisms are well understood.

Because of the high anisotropy, the superconducting properties are quite different from those of the conventional superconductors. Models of flux dynamics have to be extended to explain observations like giant flux creep in the new superconductors. In the following chapters, materials science aspects and some important physical properties of the ceramic superconductors will be treated in more detail.

The HTSCs are oxide compounds containing several metals. Their crystal structure is closely related to the perovskite structure (e.g. $SrTiO_3$). All HTSC compounds, with few exceptions, are copper oxides. Only the rare-earth (RE) compounds RE–Ba–Cu–O and the bismuth and thallium compounds are considered to be relevant for applications. Material development is most advanced for $YBa_2Cu_3O_{7-\delta}$ and the bismuth compounds $Bi_2Sr_2CaCu_2O_8$ and $(Pb,Bi)_2Sr_2Ca_2Cu_3O_{10}$. Therefore this treatment will focus on these three compounds.

Preparation of these compounds is a challenge. The quaternary system $YBa_2Cu_3O_{7-\delta}$ has a rather simple crystal structure and can be depicted in a ternary phase diagram. The five-component system $Bi_2Sr_2CaCu_2O_8$ and the lead-stabilized $(Pb,Bi)_2Sr_2Ca_2Cu_3O_{10}$ with six components, however, are more complicated.

3.1 Crystal Structure

3.1.1 The Rare-Earth Compounds

The compound $YBa_2Cu_3O_{7-\delta}$, denoted in the following by the shorthand YBCO or Y-123, has the simplest crystal structure among the superconducting cuprates. Figure 3.1 shows the tetragonal and orthorhombic phases according to [42]. As can be seen, the copper atom has an octahedral oxygen neighborhood, which is a signature of the perovskite structure. An important characteristic of all HTSC compounds is two or more adjacent CuO_2 planes, in the case of Y-123 separated by an yttrium atom. The rare-earth compounds like Y-123 or Nd-123 are the only representatives of the HTSCs which in addition have Cu–O chains in between the planes. The tetragonal-to-orthorhombic transition is coupled to an increase of oxygen content in these chains. This oxygen content is crucial for the superconducting properties of this material. For $x = 1$, i.e. in the tetragonal phase, YBCO is insulating. Below $x = 0.5$ the material becomes superconducting but with a low transition temperature. With decreasing x, T_c rises and reaches its maximum of 94 K at $x = 0.15$. This shows the important role of the oxygen vacancies in the superconducting properties of this material.

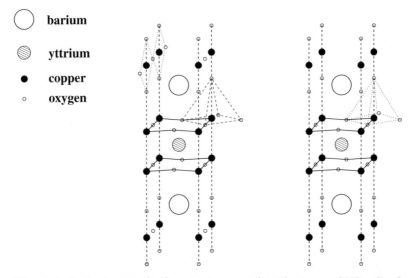

Fig. 3.1. Orthorhombic (*left*) and tetragonal (*right*) phases of $YBa_2Cu_3O_{7-\delta}$

The role of yttrium in the lattice is less important. This was demonstrated by exchanging it with rare-earth elements from the lanthanides such as Nd, Eu, Gd, Yb, Dy or Sm, which did not change significantly the superconducting transition temperature. However, the ionic radius is important as, e.g., Nd can exchange lattice sites with Ba and thus hamper the phase-pure preparation of Nd-123.

Owing to the orthorhombic crystal structure, the a and b axes are slightly different in length. The exact lengths are given in Table 3.1. This leads to twin formation in YBCO. These twins can be made visible as a tweed structure under polarized light. Twin formation, oxygen vacancies and lattice distortion due to interchange of Nd and Ba can serve as possible pinning centers for vortices as will be discussed below (Sect. 4.3).

3.1.2 The Bismuth Compounds

The superconductors of the bismuth family consist of five constituents and sometimes lead in addition. Their crystal structure is more complicated than that of the RE–Ba–Cu–O compounds as there are intercalated layers of Bi–O between the CuO_2 planes. The general chemical formula for the homologous series of the bismuth compounds is $Bi_2Sr_2Ca_{n-1}Cu_nO_{4+2n-1}$, where n denotes the number of CuO_2 planes. Figure 3.2 shows the crystal structure of the compounds for $n = 1, 2, 3$. The indices of the formula are used to construct the acronyms of the compounds, e.g., Bi-2201, Bi-2212 and Bi-2223 [179, 196]. The transition temperature rises with increasing number of CuO_2 planes up to $n = 3$, which gives the highest T_c of 110 K. Optimally oxygen-doped Bi-2212 has $T_c = 92$ K and the transition temperature of Bi-2201 lies at 10 K. This implies that only the first two compounds are relevant for applications.

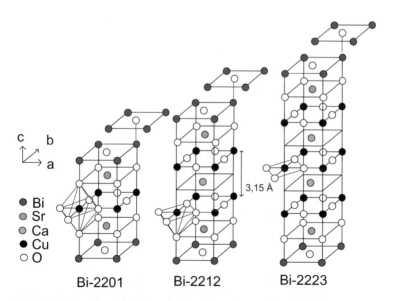

Fig. 3.2. Homologous series of bismuth superconducting compounds for $n = 1, 2, 3$

The crystal structure is tetragonal so there is no twin formation in these compounds. However, owing to incorrect stacking of planes, there can be intergrowth of Bi-2212 and Bi-2223 within one single grain. This has been demonstrated in transmission electron microscopy (TEM) studies [78].

3.1.3 Lattice Parameters, Anisotropy and Microstructure

Important quantities, namely the lengths of the crystal axes, the transition temperature and the anisotropy parameter Γ at optimum oxygen content are given in Table 3.1 for the three relevant HTSC compounds. The layered

Table 3.1. Lengths of crystal axes, transition temperature and anisotropy parameter at optimum oxygen content for the three relevant HTSC compounds

HTSC	a (nm)	b (nm)	c (nm)	T_c (K)	Γ
Y-123	0.38	0.39	1.2	94	5
Bi-2212	0.54	a = b	3.09	92	10
Bi-2223	0.54	a = b	3.70	110	12

crystal structure causes highly anisotropic electrical properties. The electrical conductivity along the ab plane is much higher than along the c axis. This is quantified by different effective electron masses in the different crystal directions. So the anisotropy parameter is defined as $\Gamma = \sqrt{m_c/m_{ab}}$.

The anisotropic crystal structure of the HTSC compounds also leads to anisotropic crystal growth. The growth velocity within the ab planes is much larger than in the c direction. The microcrystals grow as thin platelets several hundred microns in size along the ab planes and less than ten microns in size in the c direction. In a polycrystalline material these platelets are randomly stacked and some effort has to be undertaken to induce texture into the ceramic. Commonly used methods to obtain textured growth are applying a temperature gradient during thermal processing and uniaxial pressure before or between thermal cycles. Nevertheless, twin formation in YBCO cannot be avoided. The twins can only be removed afterwards by applying high pressure in order to align the a and b axes. The typical tweed structure produced by twins in polarized light is shown in Fig. 3.3. Large monolithic mixed crystals of Y-123 with Y-211 inclusions are grown by melt texturing, which will be described in more detail in Sect. 3.3. These bulk samples consist of either only one grain or a few grains, often with high-angle grain boundaries in between. Figure 3.4 exhibits such a grain boundary and the typical platelet structure of the YBCO grains. In Fig. 3.5 the very-fine-grained microstructure of a sintered pellet of the same material is shown for comparison.

In Bi-2223 melt growth is not possible as the material melts irreversibly when heated above the melting temperature. Thus for preparation of tapes

3.1 Crystal Structure 21

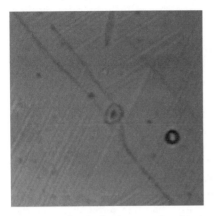

Fig. 3.3. Optical micrograph of twins in melt-textured $YBa_2Cu_3O_{7-\delta}$

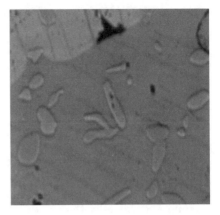

Fig. 3.4. Optical micrograph of platelets and grain boundaries in melt-grown $YBa_2Cu_3O_{7-\delta}$

from this superconductor, uniaxial pressing or rolling is used to induce texture into the ceramic core. This leads to stacking of the platelet-like microcrystals, with a preferred orientation of the c axes normal to the tape plane. Figure 3.6 shows the ceramic core of a Bi-2223 tape taken by scanning electron microscopy (SEM). The preparation process will be discussed in more detail in Sect. 3.4. The microstructure shows the typical laminar growth. The platelets are very thin and the stacking is imperfect. The typical size of the platelets is 10 µm in the ab planes and 1–3 µm in the c direction. There are different kinds of grain boundaries, and large voids between grains, which clearly manifests the need for optimization of the microstructure in this material.

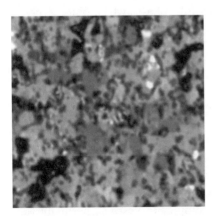

Fig. 3.5. Optical micrograph of a sintered pellet of $YBa_2Cu_3O_{7-\delta}$

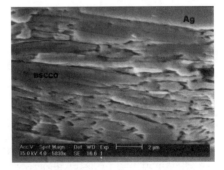

Fig. 3.6. SEM micrograph of the ceramic core of a Bi-2223 tape

3.2 Phase Diagrams and Phase Formation

As already emphasized, HTSCs are multicomponent systems. Consequently phase formation in these materials is very complex. Nevertheless, phase formation is crucial to understand, as it is closely connected to the microstructure, which in turn is responsible for the mechanical and electrical properties. This leads us to the concept of phase diagrams, which describe the equilibrium between different phases. A phase diagram contains all the information about possible phases and their coexistence at a given temperature and/or pressure at thermodynamic equilibrium. It does not contain information about the reaction kinetics, which is very important too. To understand the formation of phases and microstructure in the HTSCs one has firstly to know the phase diagram and secondly the reaction kinetics. But the phase diagram contains the basic information. For a clear presentation of the subject, basic terms, concepts and rules starting from equilibrium thermodynamics will be given in the beginning of this section.

3.2.1 Basic Concepts

To start with, some often-used terms will be introduced.

- *Component.* Components are the elements and/or compounds which are considered as the basis for phase formation
- *System.* A system is the series of possible phases which can be formed by a given set of components, for example the iron–carbon system.
- *Phase.* Every homogeneous portion of a system with uniform chemical and physical characteristics is a phase. Systems consisting of only one phase are called *homogeneous*, multiphase systems are called *heterogeneous*. Most real systems are heterogeneous.
- *Phase boundary.* A phase boundary separates phases which differ either in chemical or physical characteristics or in both. For example ice, water and water vapor are phases with different physical properties, while an oversaturated solution of water and salt contains phases that differ chemically and physically. One is a solution of H_2O, and NaCl, and liquid; the other is pure NaCl, and solid.
- *Phase equilibrium.* In heterogeneous systems several phases can coexist in time. The condition for this coexistence is given by thermodynamics. A well-known phase equilibrium exists at the triple point of H_2O, where water, ice and vapor can coexist.

The concept of phase equilibrium is important for the discussion of phase diagrams. Therefore the basic thermodynamics will be addressed briefly.

Chemical reactions in alloys or multicomponent systems normally take place in a furnace under ambient or controlled pressure. Thus T and p are the relevant external variables. To discuss equilibrium in such a system, we

3.2 Phase Diagrams and Phase Formation

have to consider, therefore, the thermodynamic potential which is connected to these variables, the *free enthalpy* or *Gibbs free energy*

$$G = E - TS + pV, \qquad (3.1)$$

where E is the internal energy of the system, S the entropy and V the volume. In equilibrium at constant pressure, G has a minimum:

$$\partial G_{T,V} = \mathrm{d}E - T\mathrm{d}S + p\,\mathrm{d}V = 0. \qquad (3.2)$$

In addition we have to take into account the fact that material can be exchanged between phases. A change of the free enthalpy at constant temperature and pressure is the consequence. If $\mathrm{d}n_i$ is the number of atoms or molecules of the component i added to a phase, G changes proportionally to $\mathrm{d}n_i$ by $\mu_i\,\mathrm{d}n_i$, where μ_i is the *chemical potential* of the component i. When several components are added, it follows from the first and second Laws of thermodynamics that

$$\mathrm{d}G = V\mathrm{d}p - S\mathrm{d}T + \sum_i \mu_i\,\mathrm{d}n_i. \qquad (3.3)$$

Then μ_i can be defined as

$$\mu_i = \left.\frac{\partial G}{\partial n_i}\right|_{T,p,n_j}. \qquad (3.4)$$

From the equilibrium condition (3.2) it now follows that

$$\sum_i \mu_i\,\mathrm{d}n_i = 0. \qquad (3.5)$$

If there are two phases α and β present in a system with components A and B at a certain temperature and pressure, the following condition also has to be valid:

$$\sum_{i=\mathrm{A,B}} \mu_i^\alpha \mathrm{d}n_i^\alpha + \sum_{i=\mathrm{A,B}} \mu_i^\beta \mathrm{d}n_i^\beta = 0. \qquad (3.6)$$

With $\mathrm{d}n_i^\alpha = \mathrm{d}n_i^\beta$ it follows that

$$(\mu_\mathrm{A}^\alpha - \mu_\mathrm{A}^\beta)\mathrm{d}n_\mathrm{A} + (\mu_\mathrm{B}^\alpha - \mu_\mathrm{B}^\beta)\mathrm{d}n_\mathrm{B} = 0. \qquad (3.7)$$

As the phases α and β should be in equilibrium concerning exchange of A and B, we finally obtain

$$\mu_\mathrm{A}^\alpha = \mu_\mathrm{A}^\beta \quad \text{and} \quad \mu_\mathrm{B}^\alpha = \mu_\mathrm{B}^\beta, \qquad (3.8)$$

meaning that the chemical potentials of two phases in equilibrium have to be equal for all components.

A more general consideration leads to the *Gibbs phase rule*. Let us consider a system with C components and P phases in equilibrium. A total of C concentrational variables is needed to describe one phase, and thus PC variables are needed to describe the system. There are P relations between the concentrational variables like

$$n_1 + n_2 + n_3 + \ldots + n_c = 1, \tag{3.9}$$

and for each component there exists a set of equations of the kind (3.8). There are $P-1$ equations in each such set, for C components, i.e. $C(P-1)$ equations. The number of independent variables or degrees of freedom F is now obtained by finding the difference between the total number of equations and the number of variables in the system:

$$F = PC + 2 - P - C(P-1),$$
$$F = C - P + 2. \tag{3.10}$$

This is the general expression for the number of degrees of freedom in a system of C components and P phases. Equation (3.10) is called the Gibbs phase rule.

During processing one degree of freedom is normally kept constant, for example pressure. In a one-component system there can be at maximum $F = 3 - P$, which means three phases in equilibrium. The system then no longer has any degrees of freedom (e.g., the triple point of water).

A more detailed discussion of phase diagram calculations and their thermodynamics is given by Haasen [82] and Castellan [35].

3.2.2 Binary Phase Diagrams

The concept introduced above will now be applied to example systems. As the first example, a fictitious binary phase diagram will be discussed at constant pressure (Fig. 3.7). The concentrations of the two components A and B are given on the x axis, while the y axis is the temperature. At point A the mixture contains 100% A, and at point B 100% B. With iron and carbon as the two components we would get the very important phase diagram of steel. Above the temperature T_1 the system is completely liquid. The boundary line below which one or two components become solid is called the *liquidus* line. Below this line there are several regimes in which either component A condenses into the solid phase α or component B condenses into the solid phase β. The solid phases are still mixed with the liquid of the other component. Furthermore, in this example, there exists a compound A_3B at a concentration of 75% A and 25% B on a line below T_2.

If a reaction starts at point L1 in the liquid region it runs on a vertical line until it reaches the liquidus line. There component B starts to precipitate as solid and the concentration in the mixture β + liquid is shifted. The reaction now runs along the liquidus line until it reaches point E, where also A_3B

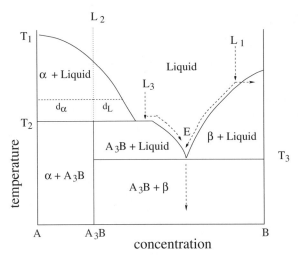

Fig. 3.7. Fictitious binary phase diagram for the components A and B

starts to precipitate as a solid phase. Now the two solid phases exist as a conglomerate. Point E is called the *eutectic* point and the two phases form a eutectic. This eutectic will also be formed when starting from point L_3.

The concentrations of the components in one phase are determined by the *lever rule*. Starting from point L_2, the reaction meets the liquidus line A and starts to condense into the solid phase α. The concentration n_α and the liquid concentration n_L at a certain temperature are then given by

$$\frac{n_L}{n_\alpha} = \frac{d_L}{d_\alpha}. \tag{3.11}$$

A more complicated case appears on heating the compound A_3B to T_2. The vertical line ends here. This means the compound decomposes into the liquid and the solid phase α; it melts *incongruently*. Such a reaction is called a *peritectic* reaction and it follows the equation

$$A_3B \rightarrow \alpha + \text{liquid}. \tag{3.12}$$

Starting from point L_2 with an exact mixture of 1:3 A:B, upon cooling component A starts to solidify first. At temperature T_3 the remaining liquid consists of 50:50 A:B. That means that A_3B can only be formed by incorporating A from the solid phase α. This has to take place via solid-state diffusion and thus at this point the reaction becomes very slow. That this can be a problem in real systems, especially in the phase formation of HTSC materials, will become clear below.

3.2.3 Ternary Phase Diagrams

The two-component system with only one degree of freedom (temperature) could be depicted quite easily in two dimensions. For the representation of a three-component system the third dimension has to be included. We then obtain a ternary phase diagram, with liquidus surfaces instead of liquidus lines. The base of such a ternary phase diagram is depicted in Fig. 3.8 as a triangle with the components A, B and C on the corners. Their concentrations are drawn as a triangular lattice.

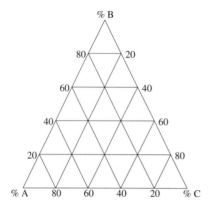
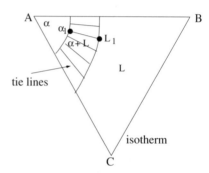

Fig. 3.8. Base of a ternary phase diagram for the components A, B and C

Fig. 3.9. Isothermal slice of a ternary phase diagram

Taking into account the temperature as the z axis, we get a three-dimensional object. Such a fictitious ternary phase diagram is shown in Fig. 3.10. In this example the binary subsystems of the ternary systems are all eutectic. In a ternary system many more compositional variables have to be considered. As a consequence the cooling route is very important and the end product depends strongly on the starting composition as well as on the solidification route. Let us start at point L in the liquid phase. Upon cooling, the liquidus surface of the solid phase α is crossed and this phase condenses, enriched in component A. Further cooling takes place along the liquidus surface and the remaining liquid gets richer in B and C, while more and more solid-phase α precipitates until the eutectic trough is reached. At this trough, mixed crystals rich in B start to solidify. The concentration of the melt moves along the eutectic trough to the ternary eutectic point E and finally solidifies into a ternary eutectic.

At any given temperature an isothermal slice can be cut from the ternary phase diagram (Fig. 3.9). Another important point can be made clear here. The liquidus lines of the two phases α and L are connected by *tie lines*, which show the different compositions of liquid and solid phase in equilibrium at that given temperature. Tie lines correspond to the lines of constant temperature in the two-phase region of a binary phase diagram. Thus the lever rule (3.11) can be used to determine the concentrations of the various phases.

The ternary phase diagram contains rather complex information. In order to concentrate on a certain region of interest, a vertical cut section can be taken at a relevant concentration. The result is a pseudobinary phase diagram. An example is given in Fig. 3.10, where the section has been taken at 80% A/20% B – 60% C/40% B. The disadvantage of this simplification is that important information may be lost in the process.

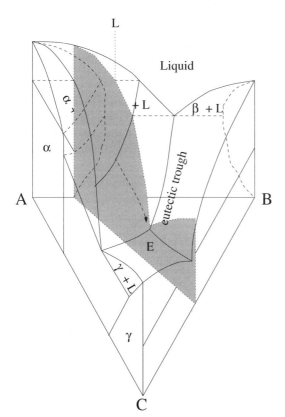

Fig. 3.10. Binary subsystem of a ternary phase diagram

We can consider Fig. 3.10 further. Starting with the stoichiometric composition of point L, the solidification reaction ends at point E with the corresponding composition and second phases. However, phase diagrams are evaluated under the assumption of thermal equilibrium. Thus the reaction described is also considered to take place under thermal-equilibrium conditions. But in a real reaction this is not necessarily the case. Some reactions are very slow and consequently the cooling process may be too fast to thermally equilibrate the system. This shows that the reaction kinetics are also a

28   3. Fundamentals of Material Processing

very important factor, not considered in phase diagrams. Therefore a preparation process needs to be labeled with all important details, not only the concentration and phase composition of the precursor material and the reaction temperature, but also the reaction time as well as the heating and cooling rates.

3.2.4 The Phase Diagram of Y–Ba–Cu–O

With the basic concepts in mind let us now consider the real systems, starting with Y–Ba–Cu–O. This obviously is a quaternary system. Three dimensions were necessary to represent a ternary system. Systems with more than three components cannot be depicted in a diagram; they have to be reduced. The system Y–Ba–Cu–O is reduced by considering the metal oxides as components of a corresponding ternary system. That is, yttrium oxide (Y_2O_3), copper oxide (CuO) and barium oxide (BaO) or barium carbonate ($BaCO_3$)[1] are used as the corners of the base of a ternary phase diagram. This base is shown in Fig. 3.11 [1]. The shaded area is depicted as a ternary phase diagram in Fig. 3.12 [180, 269]. The relevant corner points are CuO, Y-123 and Y-211.
One important feature of this phase diagram is that again the reaction route

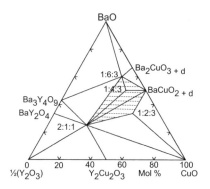 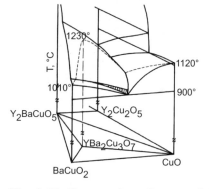

Fig. 3.11. Base of the Y_2O_3–BaO–CuO phase diagram (after [1])

Fig. 3.12. Ternary phase diagram of the shaded area in fig. 3.11 containing Y-123 and Y-211 [180]

is most important. Let the reaction start at a temperature above 1230°C with the exact stoichiometric composition of Y-123, as indicated in Fig. 3.12. Despite starting with the exact stoichiometric composition, the end product will mainly consist of Y-211, $BaCuO_3$ and CuO instead of Y-123. The reason is that the stoichiometric composition Y-123 lies within the *primary phase field* of Y-211. A comparable situation occurred in Fig. 3.10. Upon cooling

---

[1] In fact BaO is necessary for the reaction, but it reacts quite strongly with ambient air to form $BaCO_3$.

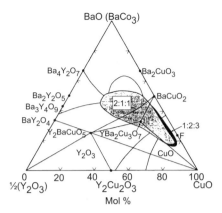
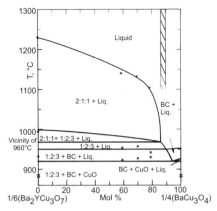

Fig. 3.13. Primary phase field of Y-211 and Y-123 [247]

Fig. 3.14. Pseudobinary section of the phase region of $YBa_2Cu_3O_{7-\delta}$ showing the small stability region of $YBa_2Cu_3O_{7-\delta}$ as the *shaded area* [180]

from point L the solid phase α was the first to solidify. Point L thus lies within the primary phase field of α. Accordingly, in Fig. 3.12 Y-211 is the first phase to solidify. In Fig. 3.13 this is made clear by superimposing the primary phase field of Y-211 as a shaded area onto the base of Fig. 3.11 [247]. The primary phase region of Y-123 is depicted as a black area in the same figure, showing that it is comparably small.

It is quite clear that, because of these peculiarities, it is not easy to produce phase-pure Y-123. In fact one has to work off-stoichiometric in the copper-rich region. In this region, however, according to the lever rule (3.11), only very little Y-123 will precipitate, revealing the major problem in growing single crystals of this material. One has to get rid of the copper-rich phase by decantation, and the small single crystals (some 100 μm in size) have to be separated from the unwanted second phases.

In Fig. 3.14 a pseudobinary phase diagram is drawn from the point Y-123 to the point F on the Ba–CuO concentration line. This shows how small the phase-pure region of Y-123 is in temperature range (around 960°C) as well as in concentration. But to make the best out of the given situation, melt-textured growth processes have been applied to obtain large mixed crystals of Y-123 with Y-211 inclusions that serve as efficient pinning centers. This is made clear by a closer look into the pseudobinary phase diagram (Fig. 3.15, from [237]). This phase diagram is drawn along a tie line between Y-211 and Y-123 from the Y_2O_3–BaO to the CuO–BaO concentration line. There are two peritectic points and one eutectic point. The two peritectic reactions are

$$122 \rightarrow 211 + \text{liquid}, \quad 1015°C < T_p < 1300°C; \tag{3.13}$$
$$211 + \text{liquid} \rightarrow Y_2O_3 + \text{liquid}, \quad 1300°C < T_p < 1500°C. \tag{3.14}$$

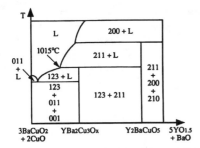

Fig. 3.15. Pseudobinary phase diagram of Y–Ba–Cu–O drawn between Y-211 and Y-123 from the Y_2O_3–BaO to the CuO–BaO concentration line [237]

These reactions are used to produce large monolithic ceramics of Y-123 with Y-211 inclusions. A detailed description of the melt-texturing process will be given below.

3.2.5 The Phase Diagram of Bi(Pb)–Sr–Ca–Cu–O

The bismuth superconductors consist of five or, if lead-stabilized, even six components. This means that the phase diagram in this case has to be reduced, too, by taking the metal oxides as components and by incorporating lead into the bismuth oxide. As a result we obtain the quaternary phase diagram with the corner points Bi_2O_3, SrO, CaO and CuO, which is depicted in Fig. 3.16 [182]. Several phase-pure regions are depicted in this diagram, which is valid at 850°C in air. In the middle of the diagram the Bi-2201

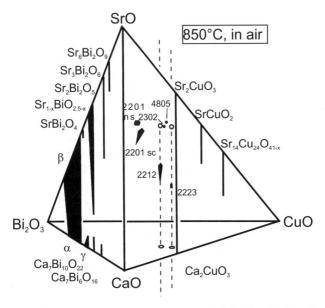

Fig. 3.16. Quaternary phase diagram of Bi–Sr–Ca–Cu–O at 850°C in air [182]

phase (also called the Raveau phase), the Bi-2212 phase and the very small region of the Bi-2223 phase can be found. Again regions of interest have to be presented as sections of the original phase diagram, taking into account the loss of information. In Fig. 3.17 a pseudoternary section containing the Bi-2212 and Bi-2223 phases is shown.

The Bi-2212 phase exists phase-pure over a certain region; that means that the Bi, Sr, Ca and oxygen contents can vary around the ideal stoichiometry of $Bi_2Sr_2CaCu_2O_8$. The role of the oxygen content will be discussed separately. As in YBCO the primary phase field of Bi-2212 does not contain the ideal composition. This implies that a stoichiometric sample almost always contains second phases as impurities. The most common phases are Ca_2CuO_3, $Sr_{14-x}Ca_xCu_{24}O_{40-y}$ with $x \simeq 7$ (14:24 alkaline-earth cuprate) and $Sr_{14-x}Ca_xCu_{24}O_{40-y}$ (copper-free phase) with $x \simeq 1$.

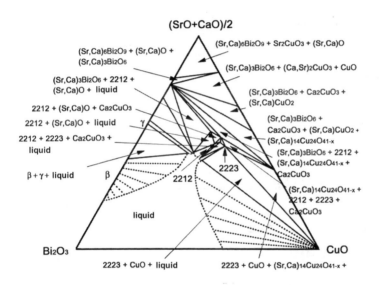

Fig. 3.17. Pseudoternary section of the Bi–Sr–Ca–Cu–O phase diagram at 850°C in air [182]

The single-phase region of $(Pb,Bi)_2Sr_2Ca_2Cu_3O_{10}$ is very small. As in the other two cases, the stoichiometric compound does not lie in the primary phase field. Through substitution of bismuth by lead, the single-phase region can be enhanced and phase stability improved. The stability region is shifted from 840°C – 890°C to 830°C – 860°C. Figure 3.18 shows a pseudobinary phase diagram based on the tie line between Bi-2201, Bi-2212 and Bi-2223. The presentation reveals some of the phase equilibria and stability regions. With increasing n in $Bi_2Sr_2Ca_{n-1}Cu_nO_{4+2n-1}$ the melting temperature and stability region decrease. For $n = 2$ Bi-2212 is stable up to 895°C, where

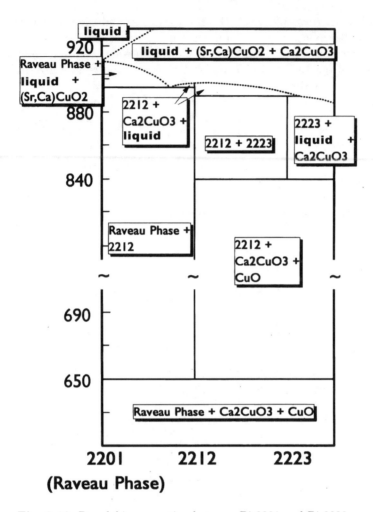

Fig. 3.18. Pseudobinary section between Bi-2201 and Bi-2223

it melts incongruently. From the glass phase, Bi-2212 can be obtained at a temperature above 650°C. The Bi-2223 phase has a very small phase stability region of width 50°C below 890°C. The Bi-2201 phase is very stable from below 650°C up to 910°C. The exact lower phase boundary is not known. Bi-2223 is obtained very slowly from a solid-state reaction, while Bi-2212 can be melt grown. The Bi-2223 crystallization can be accelerated if there is a liquid phase present in a multiphase sample. It is assumed that this is the role of the lead substitution, to produce liquid calcium plumbates to support the reaction.

The equations for the equilibrium reactions of the two- and three-layer bismuth compounds are

$$\text{Bi-2201} + 1/2\ \text{Ca}_2\text{CuO}_3 + 1/2\ \text{CuO} \rightarrow \text{Bi-2212}\,, \tag{3.15}$$

$$\text{Bi-2212} + \text{Ca}_2\text{CuO}_3 + \text{CuO} \rightarrow \text{Bi-2223}\,. \tag{3.16}$$

There are many more coexisting phases for Bi-2212 than for Bi-2223, as can be seen from the tables in [181]. In [181, 182] Majewski discusses the phase formation in the Bi–Pb–Sr–Ca–Cu–O system in great detail.

Again it has to be mentioned that phase diagrams contain only information about phase coexistence in thermodynamic equilibrium. As in the case of YBCO, reaction kinetics play an important role in the phase formation of the bismuth compounds. An additional problem arises from the fact that carbonates are used as precursor powders, $BaCO_3$ for the preparation of $YBa_2Cu_3O_{7-\delta}$ and $SrCO_3$ or $CaCO_3$ for the preparation of the bismuth materials. Through these precursors and the ambient atmosphere carbon is introduced into the starting powders, and thus carbon is present as an additional component in the system. Furthermore, when considering conductor fabrication from the bismuth superconductors, the silver cladding has to be taken into account as a further component. Because of its practical importance the latter issue will be discussed in more detail in a later section.

3.2.6 Oxygen Content

In order to reduce the number of components and to achieve easier understanding, the phase diagrams were represented above with the metal oxides as basic components. This is a simplification because the oxygen content still plays a crucial role and has to be taken into account.

In $YBa_2Cu_3O_{7-\delta}$ ceramics oxygen can diffuse easily along grain boundaries. Within the copper oxide chain the oxygen bond is not very strong; thus adding oxygen to these chains or removing it is quite easy. The oxygen concentration can vary in the range $0 \leq \delta \leq 1$. At $\delta = 0.5$ the crystal structure changes from tetragonal to orthorhombic. Only the orthorhombic phase is superconducting. The structural phase transition shows up in the lattice constants. In Fig. 3.19 the length of the crystal axes as a function of oxygen content measured by x-ray diffraction (XRD) [38, 114] is shown.

In the orthorhombic phase the superconducting transition temperature is a function of oxygen stoichiometry, as is depicted in Fig. 3.20, where T_c and the electronic states are shown as a function of δ. There are two plateaus, at 60 K and 90 K. For $\delta \geq 0.7$ the material undergoes a metal–insulator transition. In the insulating state $YBa_2Cu_3O_{7-\delta}$ shows antiferromagnetic ordering.

The bismuth compounds tend to have an oxygen surplus. The stoichiometry can vary from 8 to 8.3 in $Bi_2Sr_2CaCu_2O_8$ and from 10 to 10.8 in $(Pb,Bi)_2Sr_2Ca_2Cu_3O_{10}$. There is also a variation of transition temperature

34   3. Fundamentals of Material Processing

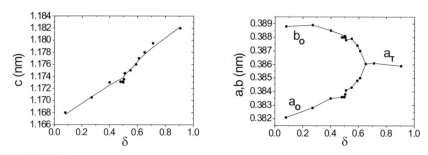

Fig. 3.19. Lattice parameters as a function of oxygen concentration δ in $YBa_2Cu_3O_{7-\delta}$ [114]

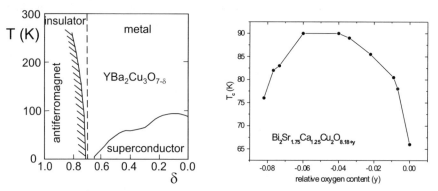

Fig. 3.20. T_c and electronic states in $YBa_2Cu_3O_{7-\delta}$ as a function of the oxygen content [36]

Fig. 3.21. T_c as a function of oxygen doping in polycrystalline Bi-2212 [197]

with oxygen stoichiometry in these materials. Optimally doped Bi-2212 has a T_c of 92 K. The transition temperature T_c as a function of oxygen content is depicted in Fig. 3.21 for polycrystalline Bi-2212 [197].

3.3 Melt Processing of $YBa_2Cu_3O_{7-\delta}$

Preparation techniques for technical high-temperature superconductors differ substantially from single-crystal growth processes. On one hand, for system design much more material is needed than can be prepared in a reasonable time by single-crystal growth. On the other hand, the perfection of single crystals may not be useful as defects are needed as pinning centers. Thus processes have been developed to produce large amounts of HTSCs for technical applications.

One of these processes is the melt-textured growth (mtg) of large, monolithic $YBa_2Cu_3O_{7-\delta}$ samples. When $YBa_2Cu_3O_{7-\delta}$ was discovered [178], for some time it could only be produced as pressed and sintered pellets.

The sintering temperature was slightly below the melting temperature of $YBa_2Cu_3O_{7-\delta}$. The critical current densities which could be achieved with these samples were disappointingly low. The reason was that in these very-fine-grained materials a large number of high-angle grain boundaries are present, which carry nearly no supercurrent as is known from bicrystal experiments [91].

3.3.1 Melt Growth Processes

All melt growth processes are based on the phase reactions described in Sect. 3.2.4. The first processing scheme, applied by Jin et al. in 1988 [113], consisted of heating a sintered pellet of stoichiometric $YBa_2Cu_3O_{7-\delta}$ above its melting point and subsequently cooling down slowly to enable recrystallization. Bulk samples prepared by this method were still very polycrystalline but their critical current densities turned out to be slightly higher than in sintered samples. In the following years this process was modified and improved by several researchers. The first improvement was achieved by Salama and coworkers [236] by introducing the liquid-phase processing (LPP) technique. As in the process used by Jin et al., a stoichiometric sintered pellet is heated above the first peritectic temperature of 1015°C. But in contrast to the first process, a dwell time is introduced at that elevated temperature in order to obtain a homogeneous melt. After the dwell time the melt is cooled rapidly to a temperature slightly above the melting temperature and then subsequently cooled very slowly to a temperature well below the melting point. This again enables slow recrystallization. The schematic temperature–time diagram is shown in Fig. 3.22.

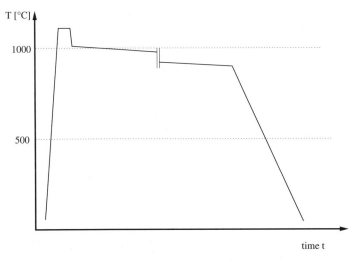

Fig. 3.22. Temperature–time diagram of the LPP process after Salama [236]

3. Fundamentals of Material Processing

The technique introduced by Murakami et al. [204] is called the melt powder melt growth (MPMG) process. Murakami et al. added a further step to the LPP process, in which the precursor powders of Y_2O_3, CuO and $BaCO_3$ are heated above the second peritectic temperature of $YBa_2Cu_3O_{7-\delta}$ (1300°C). After a short hold at this temperature the melt is quench-cooled, crushed and pressed again into pellets, which now undergo an LPP processing. This additional step provides an even more homogeneous distribution of the constituents. The schematic time–temperature diagram is shown in Fig. 3.23.

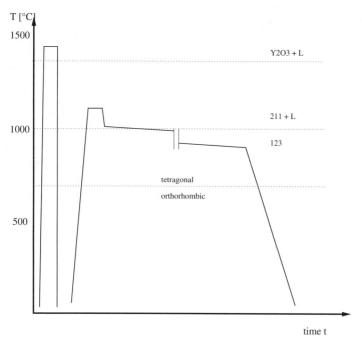

Fig. 3.23. Processing scheme of the MPMG process after Murakami et al. [204]

Further techniques use temperature gradients in the furnace to achieve directional solidification. Thus the degree of texture in the material is higher, and only low-angle grain boundaries are introduced. Ulrich and coworkers [260] used a modified Bridgman technique where a hot zone is driven through a rod-shaped sample in a multizone furnace. The traveling melting zone also leads to directional solidification.

All the processes described take advantage of the special features of the Y–Ba–Cu–O phase diagram as depicted in Fig. 3.15. As already mentioned above, the stoichiometric composition of $YBa_2Cu_3O_{7-\delta}$ lies within the primary phase field of the insulator Y_2BaCuO_5 (Y-211). As a consequence this compound is obtained first upon cooling from the melting point. At first glance this seems to be a disadvantage. But it turns out that inclusions of Y-211 within the Y-123 significantly enhance the critical current density. That means that these precipitates or, more precisely, the lattice distortions around them serve as efficient pinning centers. Owing to the short coherence length in this material it is better to have very small submicron-sized precipitates which are homogeneously distributed throughout the sample. The goal of all melt growth processes therefore is to prepare large, monodomain mixed crystals of Y-123 and Y-211 where the Y-211 inclusions are of submicron size and finely dispersed in order to achieve high critical current densities.

A model for the growth of Y-123 with Y-211 inclusions was proposed by Cima et al. [37] and will be described using Fig. 3.24. When the material is heated above the second peritectic temperature, only Y_2O_3 remains as a solid; thus yttrium can be finely dispersed in the melt. Upon cooling below this temperature, Y-211 is obtained as a solid phase. Upon further cooling below the first peritectic temperature, Y-123 recrystallizes at the solid–liquid interface. Thus Y-123 crystals grow around the Y-211 particles as depicted in Fig. 3.24. In order to form the Y-123, yttrium from the Y-211 phase has to be incorporated. This can only happen through solid-state diffusion, which is a very slow process. Once the Y-211 particles are completely surrounded by Y-123, material from the interior of the inclusion is no longer transported into the melt. That is, if the preparation is started with a stoichiometric composition of the precursors there is a lack of yttrium in the solid Y-123 phase and thus $BaCuO_3$ remains as a second phase, preferentially at grain boundaries. Hence an off-stoichiometric precursor composition is used with up to 50% yttrium surplus. In order to reduce the size of the Y-211 inclusions, addition of platinum oxide or cerium oxide has proved to be useful [121].

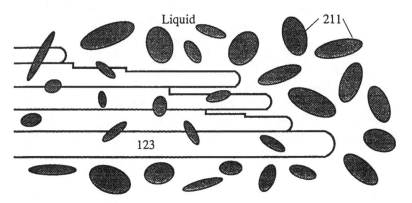

Fig. 3.24. Growth model for Y-123/Y-211 mixed crystals [237]

3.3.2 Top-Seeding Technique

For technical applications monolithic samples consisting of only one or a few well-connected grains are desirable. One method to achieve monodomain crystals is the aforementioned Bridgman technique, which, however, is fairly expensive. A breakthrough in the inexpensive growth of large amounts of high-quality ceramic YBCO samples was reached when an old crystal growth technique was applied: using a seed crystal [157]. In this top-seeding technique RE-123 crystals with a higher melting temperature than Y-123, like Sm-123 or Nd-123, are put on top of the precursor pellet. These crystals are about one by one millimeter wide and 100 μm thick and need not be of very high quality. During recrystallization they serve as nucleation centers where crystallization starts, and directional solidification occurs from the seed crystal. For large samples more than one seed crystal is used. With this method it is possible to grow large amounts of monodomain samples without heating above the second peritectic temperature. Various institutes around the world, such as the Institut für Physikalische Hochtechnologie (IPHT) in Jena [59], the Forschungszentrum Karlsruhe (FZK) [115] and the Texas Center for Superconductivity at the University of Houston (TSCUH) [284], are producing large amounts of single-domain $YBa_2Cu_3O_{7-\delta}$ samples of various shapes.

3.4 Tape and Wire Fabrication

Heike Kammerlingh Onnes [118] dreamed the dream of superconducting magnets and current transport without losses. It took 50 years before his findings led to the first commercially available superconducting NbTi magnet. Lossless energy transportation still remains a dream. But with the arrival of HTSCs this dream became more realistic. Therefore, from the very beginning of the material development of HTSCs, materials scientists tried to make conductors out of these materials. The first trials to produce conductors out of $YBa_2Cu_3O_{7-\delta}$ were without success owing to the already mentioned grain boundary problem. Hence conductor development concentrated on the preparation of BSCCO wires and tapes. Only recently has a second generation of YBCO coated conductors led to promising results. In the following, different preparation techniques will be addressed and some will be described in more detail.

3.4.1 Conductor Fabrication Processes

The main problem of the HTSCs is that they are ceramic materials and thus cannot be directly deformed into a wire like the metallic superconductor NbTi. There must be a matrix as a carrier or flux for the brittle material, similar to that used for Nb_3Sn. Copper is used as a matrix for conventional superconductors but it is a bad choice for the HTSCs because it reacts strongly with these materials, as they all are cuprates themselves. Silver is adequate but fairly expensive; nevertheless it remains the only possibility, apart from some of its alloys. Two principal arrangements of conductor formation have been and are still being pursued, dip coating of silver tapes and the powder-in-tube (PIT) technique. All other techniques are more or less variations of these.

Three main coating techniques have been reported: electrophoresis, dip coating and the doctor blade technique. The electrophoresis technique takes advantage of the anisotropy of the precursor material to deposit thick films on a conducting silver substrate in an electric field [89, 98, 267]. The superconductor is then formed by thermal processing. The other two techniques use a solution of precursor powder in an organic solvent. Either the silver tape is dip coated by dipping it into the solution or the ink is pasted onto the tape with a doctor blade. In all cases there has to be a thermal treatment afterwards. These thick-film coating techniques are cheap and can easily be upgraded to continuous coating. In particular, in Bi-2212 tape production, dip coating is still used [60, 86, 199] as this material does not need intermediate mechanical treatment. However, critical current densities in Bi-2212 dip-coated tapes can be significantly improved by introducing an intermediate rolling step, as has been shown by Furukawa Electric in Japan [194] very recently. With their so-called PAIR (preanneal and intermediate rolling) process they achieved $j_c = 900\,000$ A/cm^2 at 4.2 K in zero field and still 500 000 A/cm^2 in 1 T [193], which is the world best value for Bi-2212 tapes.

Tape conductors using the Bi-2223 compound can only be produced by the PIT process as intermediate mechanical treatment is crucial. Upscaling to long-length conductor fabrication is not quite as easy with the PIT process as with the dip-coating technique. Nevertheless several companies have succeeded in producing long tapes of Bi-2223, and Bi-2212 as well. The highest critical current densities reached so far with the PIT technique are $70\,\text{kA/cm}^2$ at 77 K in zero field for Bi-2223 by American Superconductor Corp. [234] and $470\,\text{kA/cm}^2$ at 4.2 K in zero field by Hitachi [205]. An extensive review of preparation processes of BSCCO wires and tapes is given by Hellstrom [94].

3.4.2 Green-Wire Fabrication Using the PIT Process

Important stages of the PIT process are the choice and preparation of the precursor powder and sheath material, the filling and deformation of the filled tubes to green wires or tapes, and the final thermal treatment and formation of the superconductor within the cladding. The strongly two-dimensional microstructure of the BSCCO materials is very advantageous in this process. Owing to the platelet structure the microcrystals easily glide over each other during deformation, thus favoring the texture formation. This does not happen with YBCO. This is one reason why the PIT process did not work in the latter material.

There is no significant difference in the preparation of Bi-2223 and Bi-2212 green tapes so the following is valid for both materials. The PIT process with its different stages is depicted schematically in Fig. 3.25.

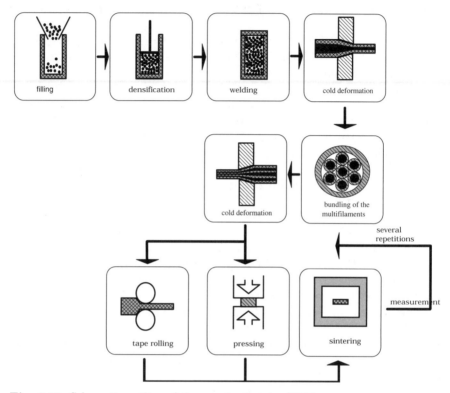

Fig. 3.25. Schematic outline of the powder-in-tube (PIT) process

The processing parameters are the following:

- precursor composition
- powder pretreatment
- sheath material
- sheath material pretreatment
- filling
- compacting of the filled material
- pretreatment of filled tubes
- deformation
- stacking to multifilaments
- thermal or thermo-mechanical treatment.

Precursor Powders

Fully reacted Bi-2212 is used as the precursor powder for Bi-2212 conductor fabrication, and Bi-2212 + $CaCO_3$ + CuO + SrO in the appropriate stoichiometry is the starting mixture for Bi-2223 tape preparation. The cation stoichiometry can vary. Often a slightly overstoichiometric content of Bi and Ca is used. The mean grain size of the powder is also an important factor. Grain sizes of 5 µm and lower are preferable The powder has to be carefully calcined in order to transform the carbonate into an oxide, as carbon is harmful to the thermal processing because it resides at grain boundaries in the final tapes, as could be shown [55].

Sheath Material

The sheath material is either silver, or a silver alloy such as Ag/Cu, Ag/Au or silver with a small amount of Mg. The tubes have to be cleaned very carefully as organic impurities such as oil or grease also introduce carbon into the system.

Filling and Compacting

The precursor material is filled into the sheath with mechanical agitation to achieve high powder densities. Methods such as cold and hot isostatic pressing (HIP, CIP) of the precursor powders and filling the resulting rods into the silver sheath have also been applied in order to obtain high filling densities. The highest densities, however, which can be achieved are around 40%.

Pretreatment of Filled Tubes

A heat treatment in vacuum for one or two days may follow before deformation. This step is useful for removing moisture from the compacted powder within the tube.

Mechanical Deformation

The principal deformation methods are extruding, swaging, drawing, rolling and pressing. Extrusion and swaging are necessary when starting with very-large-diameter batches in order to produce long lengths of conductor (100 m to 1 km). For smaller batches of up to 100 m length, rolling or drawing is sufficient as an initial deformation step. In order to produce very-high-j_c short samples, intermediate pressing is used during thermo-mechanical treatment. Rolling, however, is a more suitable technique for continuous production. The critical current densities of rolled samples are on average lower than those of pressed tapes. The reason is that the stress and strain distributions of the various deformation techniques are very different. Rolled samples tend to have microcracks normal to the rolling direction, which hinder current flow more efficiently than cracks parallel to the rolling direction, as produced by pressing [215]. Nevertheless cracks are produced during mechanical deformation owing to the brittle nature of the superconducting ceramic. A very detailed review of the mechanical deformation of BSCCO/Ag composites is given by Han et al. [84].

Stacking to Multifilaments

After deformation, monocore wires are restacked into silver or silver-alloy tubes to form multifilamentary conductors. The same deformation process has to be performed once again.

Thermo-mechanical treatment

In the last step of thermal or thermo-mechanical treatment the superconductor is formed within the silver cladding. These processes are different for the two bismuth compounds. Because of their importance for the phase formation they will be discussed separately in the next section.

3.4.3 Phase Development of Pb,Bi–Sr–Ca–Cu–O in Silver

In Sect. 3.2 the phase diagram of the Bi–Pb–Sr–Ca–Cu–O system was discussed in some detail. The application of these phase diagrams to PIT conductors of the same material is not straightforward, as the sheath material has to be considered as a further component in the system. Also, the role of the lead addition to the Bi-2223 phase is very important for the phase formation within the silver sheath.

Phase Diagrams in the Presence of Silver and Lead

Majewski [182] studied the influence of silver and lead on the phase diagram of the Bi–Sr–Ca–Cu–O system. As already mentioned above, the single-phase region of Bi-2223 is enlarged when lead is added. From XRD studies it can be assumed that most of the lead is incorporated into the compound at Bi-sites. However, the lead solubility is strongly temperature-dependent. This can be seen from Fig. 3.26, where a section through the single-phase region of Bi-2223 is drawn as a function of temperature and lead content. The lead solubility increases with temperature up to 850°C and decreases again above that temperature. A lot of second phases are present around this single-phase region, such as Bi-2212, alkaline-earth cuprates, the 14:24 phase and a phase called 451 by Majewski, which is a lead-rich phase and is called 3221 or 3321

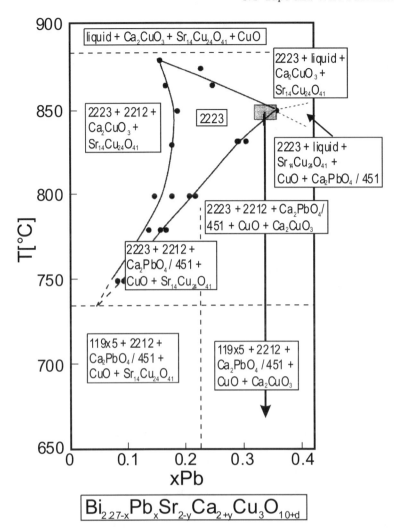

Fig. 3.26. Single-phase region of lead-stabilized Bi-2223 as a function of lead content and temperature [182]

by other authors. Its composition is roughly $Bi_{0.5}Pb_3Sr_{2.5}Ca_2CuO_x$ and it will be denoted 3221 in the following. Assuming furnace cooling from the processing temperature as indicated by the arrow in Fig. 3.26, as is normally done in conductor fabrication, it is clear from the phase diagram that second phases cannot be avoided. The 3221 phase forms and decomposes with increasing temperature. The lead is incorporated into the Bi-2212 phase. This makes the lead solubility a serious fundamental problem in wire fabrication.

The silver sheath of a PIT conductor adds one more component to the multiphase system. Two facts are fairly well known. The first is that silver

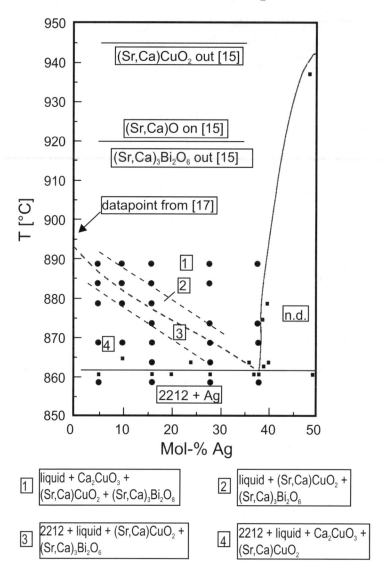

Fig. 3.27. Pseudobinary phase diagram of Bi-2212–Ag as a function of temperature [182]

is dissolved in Bi-2212 and Bi-2223; the second is that the silver addition significantly lowers all reaction temperatures, especially the melting points of both phases, by about 20°C. The phase diagram was studied only for Bi-2212–Ag by Majewski [182] and McCallum et al. [190]. Majewski's results are depicted in Fig. 3.27.

From these studies it appears that silver also influences the lead solubility. On one hand it reduces the temperature of maximum lead content from 850°C to 835°C and on the other hand it also reduces the amount of lead which can be dissolved in Bi-2212.

All the phase diagram studies were performed using mixed precursor powders pressed to pellets and sintered. What is actually going on during phase formation in a silver sheath can be concluded from these results only within certain limitations. The silver is normally homogeneously mixed with the precursor in the phase diagram studies, while in a silver-sheathed tape the silver-BSCCO interface may have a different concentration from the interior of the ceramic core. Thus the phase formation within an Ag-cladded tape has to be considered separately.

Phase Formation within the Silver Sheath

The first processing steps necessary for Bi-2223 formation are quite well established, for example by experiments performed by Grivel and Flükiger [79], and have been reviewed recently by Flükiger et al. [55]. According to these findings the formation of Bi-2223 is a two-stage process. The composition of the starting powder is chosen as stoichiometric Bi-2212 together with other phases (Ca_2PbO_4 and CuO) yielding the appropriate stoichiometry for Bi-2223. In the first, very fast reaction lead is incorporated into the Bi-2212, leading to nonstoichiometric Bi,Pb-2212 and second phases. The second reaction, yielding Bi,Pb-2223 from these phases, is very slow. In detail, the following happens:

1. Ca_2PbO_4 and CuO are consumed within the first few hours of the reaction (up to 50 h yields > 90% conversion). The lead is incorporated into the Bi-2212 phase.
2. The lead-doped Bi-2212 slowly decomposes into a transient liquid consisting of Bi-2201, 3221 and $(Ca,Sr)_{14}Cu_{24}O_y$ (the 14:24 phase).
3. As soon as this liquid exists in the appropriate stoichiometry it starts to form Bi-2223 until there is no longer enough liquid present.

At this point small amounts of the residual phases remain as amorphous layers along the grain boundaries, where they very efficiently block the current path. Up to this stage of the process it is useful to work at higher sintering temperatures in order to obtain large Bi-2223 grains, but at the expense of an appreciable amount of second phases like Bi-2201, which coexists with Bi-2223 in this temperature range, as found by Däumling et al. [41]. According to these authors there is a single-phase region of Bi-2223 at slightly lower temperatures. Thus at reduced sintering temperatures Bi-2201 can be converted by a solid-state reaction as it is no longer stable in this part of the phase diagram. This has been observed and the formation and decomposition of 3221 below 800°C has been shown experimentally by Wang et al. [265, 264].

After thermo-mechanical processing, the density of the ceramic core is much higher than before, leading to increased alignment of the growing crystals. Texture formation and crystal growth from the melt in the core of the tapes are still controversial. Flükiger et al. [55] conclude from their experiments that the lead-rich Bi-2212 has to decompose completely and Bi,Pb-2223 forms from the liquid by nucleation and growth. Other authors [46, 266], however, claim an intercalation process in which BiO_2 planes are intercalated into Bi-2212 to form Bi-2223 as the formation mechanism for Bi-2223. It cannot be excluded that both mechanisms are active in different stages of the processing or when starting with different phase compositions of the precursor powder. Fairly well established is the fact that the silver sheath seems to support the texture formation as the BSCCO grains are much better aligned there than in the middle of the core [283]. The thermo-mechanical treatment also leads to the formation of many well-connected grains, thus improving grain contact and consequently current-carrying capability.

Influence of the Atmosphere

The influence of the composition of the reaction atmosphere cannot be neglected. In particular, the oxygen content of the atmosphere seems to affect the reaction kinetics and temperature. The optimum reaction temperature is reduced by about 10°C in 8% oxygen as compared with air. In addition, the temperature window in which the reaction takes place is broader by about 3°C. In air the optimum temperature has to be regulated within ± 0.5°C. The atmosphere also influences the reaction kinetics. It has been shown [285, 286] that the Bi-2212-to-Bi-2223 conversion is faster in a reduced-oxygen atmosphere. Furthermore, the phase assemblage also is influenced by the oxygen content. In particular, the lead content of the Bi-2223 phase changes with atmosphere. Holesinger et al. [100] could show that more lead-rich second phases are present when processing in 10% oxygen as compared with air. These second phases seem to increase the critical current density. The best results were obtained with a variable oxygen atmosphere, i.e. increasing the oxygen pressure from 10% to 20% in subsequent heating cycles. But still the published results on the influence of the oxygen partial pressure are somewhat contradictory and some more detailed studies have to be performed.

3.4.4 YBaCuO-Coated Conductors

From all that has been discussed so far it is clear that conductor fabrication from the BSCCO compounds has one severe drawback, and that is the complex phase formation. The phase formation of the HTSC $YBa_2Cu_3O_{7-\delta}$ is much simpler as it consists of only four components, including oxygen. But, however, as already mentioned it is not possible to produce PIT conductors from this material as it does not form good grain contacts. Only highly aligned grain boundaries act as strong links. Thus good texture in this HTSC ceramic is hard to achieve. In fact only melt-grown bulk samples and epitaxial thin films of YBCO can carry high critical currents. A high-quality thin film is able to carry up to $3\,\mathrm{MA/cm^2}$.

In the past appreciable effort has been put into the development of thick films of YBCO on industrial substrates such as large-area metal parts of cylindrical shape [58] or long, flexible tapes of hastalloy or polycrystalline nickel [273]. As YBCO does not grow epitaxially on these substrates, biaxially aligned buffer layers of YSZ (yttrium-stabilized zirconia) and/or CeO_2 have to be deposited onto the metal before growing the superconductor film. In order to achieve good biaxial alignment of the buffer layers, ion-beam-assisted deposition (IBAD) was first used by Ijima et al. [108] and was subsequently adopted by other research groups [58, 273]. This technique uses an ion beam which is focused onto the substrate during deposition of YSZ and CeO_2 to inhibit nonaligned growth. The angle of the ion beam with respect to the substrate defines the preferred growth direction of the layers. Conductor lengths of up to one meter on hastalloy have been produced [57] with critical current densities of $2\,\mathrm{MA/cm^2}$ [212]. Nevertheless, this process is very slow. It takes one hour to produce one centimeter of conductor [57]. Furthermore, it is fairly expensive as it uses pulsed laser deposition (PLD) as the deposition technique for the superconductor.

In 1994 a new type of YBCO-coated conductor was introduced by Goyal et al. [76] from Oak Ridge National Laboratory (ORNL). The substrate was named RABiTS^{TM}, which means "rolling-assisted biaxially textured substrate". All the work done so far at ORNL has been summarized by Goyal et al. in a review article [75]. This conductor type will be discussed in some further detail in the following.

Metal substrates

The idea of RABiTS^{TM} is to use metal substrates which are already biaxially textured. It is well known from the literature [70, 71, 72, 82] that most fcc metals exhibit a pronounced cube texture when they are strongly deformed by rolling and subsequently heated in vacuum. Widely used metals such as nickel, copper or aluminum exhibit this cube texture, while silver is known to be the only fcc metal which shows the more complicated brass texture after rolling and recrystallization. Nickel was chosen as the first candidate for RABiTS^{TM}. High-quality cube texture was achieved after rolling with a reduction ratio of over 90% and subsequent heating in vacuum at temperatures between 900°C and 1100°C. However, nickel is a magnetic material, and for applications in energy technology this is a big disadvantage owing to the magnetic hysteresis losses in ac fields which would be produced by the nickel subtrate. There are several nickel alloys such as constantan (Ni–Cu) which are nonmagnetic and also show cube texture upon rolling and heating [71]. Experiments with nickel alloys have been performed by several groups, using for example Ni–V

[218] or Ni–Cr, but so far the best results have been obtained on pure nickel foils [75].

Buffer Layers

With the exception of silver, pure metals cannot serve as a substrate as YBCO has to be grown in an oxygen atmosphere and the metal substrate would be destroyed by oxidation. In addition most metals react strongly with the superconductor; in particular, nickel, as a magnetic material, is known to destroy superconductivity. Therefore buffer layers have to be grown between the metal and the $YBa_2Cu_3O_{7-\delta}$ in order to prevent oxidation and interdiffusion. Not only does this buffer layer have to be a chemical barrier but its lattice parameters ought to match those of $YBa_2Cu_3O_{7-\delta}$ in order to achieve good epitaxy. In the IBAD process these buffer layers are grown on a polycrystalline substrate. If the metal substrate is already aligned itself, this alignment is induced also in the buffer layers as could be shown by the ORNL group [75]. Therefore IBAD is no longer necessary and the process becomes much faster. A schematic outline of the process as drawn by Goyal et al. [75] is shown in Fig. 3.28.

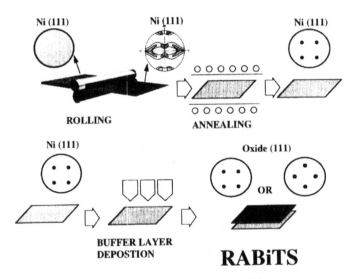

Fig. 3.28. Schematic outline of the RABiTS^{TM} fabrication process (from [75])

Up to now the most successful buffer layer structure used by the Oak Ridge group has been a multilayer of CeO_2, YSZ and CeO_2 on top of nickel, each component of the multilayer being around 500 nm thick. A noble-metal layer (Ag, Pd) between the nickel and the oxide layers proved to be a disadvantage owing to interdiffusion problems and was removed. Good out-of-plane and in-plane texture was obtained in subsequent layers, with a mosaic spread of less than 10°.

YBCO Thick Films

On top of this multilayer structure high-quality $YBa_2Cu_3O_{7-\delta}$ films of 1 to 1.4 µm thickness were grown. The highest j_c which could be reached with these YBCO RABiTS^{TM} films was 1.4×10^6 A/cm^2 over small pieces of less than 10 cm in size. The most recent results report 3 MA/cm^2 on a Ni–CeO$_2$–YSZ RABiTS^{TM} substrate [186]. But the YBCO films were only up to 250 nm thick. On longer samples of 10 cm size the critical current density reaches only 270 000 A/cm^2 [213]. The degradation of j_c in magnetic fields is very weak, even weaker at elevated fields than in YBCO films on single-crystal substrates, as can be seen from Fig. 3.29.

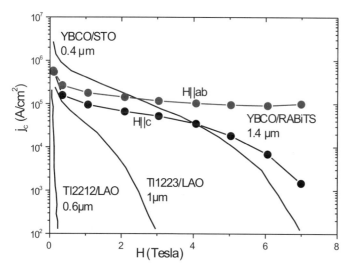

Fig. 3.29. Magnetic-field dependence of j_c of YBCO RABiTS^{TM} films at 77 K in parallel and perpendicular magnetic fields in comparison with results for YBCO films on single-crystal SrTiO$_3$ (STO) and Tl films on LaAlO$_3$ (LAO) [75]

The improved properties in magnetic fields are attributed to additional pinning sites at crystalline defects due to a nonstoichiometric local composition of the YBCO RABiTS^{TM} films [75].

It can be concluded that YBCO films on flexible metallic substrates have superior properties as compared with BSCCO conductors, especially at 77 K in high magnetic fields. Applications like superconducting magnets with liquid-nitrogen cooling will only be feasible with such a type of conductor.

However, there is still the great disadvantage of either very slow and expensive processing (IBAD) or scalability (RABiTS^{TM}). In particular, upgrading the very promising results of YBCO RABiTS^{TM} films to long lengths of conductors might be a severe problem. The reason is that these films still show cracks and thus interdiffusion remains a problem. Whether it will be possible in future to produced long-length crack-free buffer layers on well-textured metal substrates has to be investigated very carefully.

4. Physical Properties of High-T_c Superconductors

4.1 Normal-State Properties of HTSCs

The above-mentioned large structural anisotropy of the high-temperature superconductors has an influence on the physical properties in the normal as well as in the superconducting state. In the normal state the effective mass of charge carriers, the specific resistance and magnetotransport are strongly anisotropic. Only electrons from the Fermi edge contribute to the transport properties of a conductor. The electronic states reside within k space on the Fermi surface. A simple cubic metal has only one value for the Fermi energy and the Fermi surface is a sphere. The more complex the crystal structure and the higher the number of valence electrons, the more complicated the Fermi surface becomes. The Fermi surface of the fcc metal copper, for example, makes contact with the boundary at the center of the hexagonal face of the Brillouin zone (BZ) [125]. The consequence of such an anisotropic Fermi surface is anisotropy in the effective mass or resistivity.

In the normal state the HTSCs are very bad electrical conductors. Depending on their doping, e.g. by variation of the oxygen content, they are near to a metal–insulator transition. In $YBa_2Cu_3O_{7-\delta}$ the oxygen content of the copper–oxygen chains is important. In the insulating state the Cu spins are antiferromagnetically ordered. The (x, T_c) phase diagram is depicted schematically in Fig. 3.20 [36].

The determination of the Fermi surface of YBCO was difficult. In principle there are three methods of evaluation: angle-resolved photoelectron spectroscopy (ARPES), angular correlation of annihilation radiation (ACAR) and the de Haas–van Alphen effect. They all have advantages and drawbacks. Photoemission spectroscopy has to be performed in ultrahigh vacuum (UHV) and as its range is very small it can only probe a thin surface layer. That means that surface contamination and oxygen deficiency in the first few atomic layers, as happen in YBCO in UHV above 50 K, reduce the reliability of the data. The de Haas–van Alphen effect needs a very low impurity density in order to achieve a high mean free path of the electrons. If electrons are scattered by impurities or vacancies in a sample the de Haas–van Alphen period is overdamped and the results become ambiguous. Eventually it was possible to determine the Fermi surface of $YBa_2Cu_3O_{7-\delta}$ by ACAR [247].

The structure of the Fermi surface reflects the highly anisotropic material properties of the HTSCs. As most HTSC compounds have an orthorhombic crystal structure their resistivity is a tensor with three independent components ρ_{xx}, ρ_{yy} and ρ_{zz}. The c axis is defined to be parallel to the c direction and the x and y axes lie within the ab plane. As orthorhombic crystals often exhibit twinning, ρ_a and ρ_b cannot be determined independently; therefore a mean value ρ_{ab} is given. This is a good approximation as the difference between the two is normally less than a factor of two. The resistivity ρ_c, however, can be orders of magnitude higher.

The resistivity ρ_{ab} of $YBa_2Cu_3O_{7-\delta}$ shows a distinct linear behavior up to 300 K, while ρ_c is a hundred times higher and increases with decreasing temperature as is typical for a semiconductor. The anisotropy in Bi-2212 single crystals is of the order of 10^6 and shows, in principle, the same behavior as in YBCO [5].

4.2 Superconducting Properties

4.2.1 Microscopic Description

Before the HTSCs were found in 1986 [15] the research community agreed on the fact that superconductivity above 30 K was impossible according to the BCS theory. Therefore many theories on the mechanism of high-temperature superconductivity were put forward which did not consider electron–phonon coupling as the basic interaction for superconductivity as the BCS theory does. Among them there are bipolarons, anyons and marginal-Fermi-liquid theory. Recently the discussion has concentrated on either electron–phonon interaction and an extended BCS theory or spin fluctuations as the explanation for the high transition temperature and other properties of the HTSCs. This discussion is very lively among theoretical physicists and experimentalists as a decision in favor of one or other model would have far-reaching consequences for application, as will be argued below.

The explanation of high-temperature superconductivity within the framework of the theory of Bardeen, Cooper and Schrieffer (BCS theory) [12] requires the existence of a Fermi liquid, which is proven by the determination of the Fermi surface of $YBa_2Cu_3O_{7-\delta}$. As the material is close to a metal–insulator transition, however, charge carriers are much more localized than in simple metals. The relevant conduction bands are two-dimensional hybrid bands from overlapping Cu 3d and O 2p orbitals. The role of the copper oxide chains in superconductivity is marginal. It is assumed that superconductivity within these chains is introduced by the proximity effect. In the insulating state the Cu 3d orbitals are antiferromagnetically ordered, while the chain oxygen is completely absent. Even in the superconducting state this magnetic ordering is still present at small distances. It can be shown that, owing to their short-range order, spin correlations can also lead to pair formation similarly to electron–phonon interaction [74, 139]. The two mechanisms lead to different symmetries of the Cooper pairs and the superconducting wave function ψ. Electron–phonon coupling implies an orbital momentum of the Cooper pair of $l = 0$, i.e. an s state with a gap of 2Δ at the Fermi edge. The consequence of the spin fluctuation mechanism is a pairing state with $l = 2$ and thus a d-wave state. Pairing states with orbital momentum different from zero have been discussed before. Thus superfluidity in liquid He^3 and superconductivity in certain heavy-fermion systems can only be explained with p-wave pairing, which means the macroscopic superconducting wave function has orbital momentum $l = 1$ (spin triplet).

At present there is experimental evidence for both mechanisms. The energy gap is the relevant quantity under discussion. The existence of two energy gaps in $YBa_2Cu_3O_{7-\delta}$ has been shown by tunneling spectroscopy [224]. In their *two-band model* Kresin and Wolf [140] take into account these results by assuming two energy bands with two corresponding gaps. Different energy gaps also may exist in conventional superconductors, but because of the much larger coherence length these gaps will not be resolved. In HTSCs the coherence length is very short and thus more than one energy gap can be detected. Further features of the model are that phonon exchange leads to pairing and induces superconductivity into the intrinsically normal-conducting Cu–O chains. Furthermore, in conventional superconductors low-energy acoustic phonons mediate the interaction. In HTSCs – as in ferroelectric perovskites such as $KaLiTaO_3$ [162, 252] – there are softening optical phonons for small wave vectors, which contribute to the interaction with a higher energy. As this energy is inserted into the formula for the transition temperature in the BCS theory, this consideration already leads to enhanced critical temperatures. Energy states within the gap or gapless superconductivity can readily be explained within the two-band model by magnetic impurities. Many experimental findings fit into the framework of this theory. Nevertheless there is also evidence for the d-wave pairing mechanism from surface impedance measurements on thin epitaxial films [96, 97, 128, 210]. The d-wave symmetry of the order parameter leads to a spatial anisotropy of the energy gap along the Fermi surface that means that there are nodes along certain directions. As a consequence the density of states at the band edge is no longer singular but steadily goes to zero towards the middle of the band gap. Some very convincing experiments on the symmetry of the order parameter have been performed by Tsuei and coworkers on tricrystals of $Bi_2Sr_2CaCu_2O_8$ and $YBa_2Cu_3O_{7-\delta}$ [258]. The results show that, depending on the material, the wave function can have pure d-wave or a mixture of s- and d-wave symmetry. But an unambigious experimental proof of one or the other mechanism is still missing.

The consequences of a d-wave pairing mechanism would be deleterious for applications, especially in high-frequency systems made from HTSC films. Owing to occupied states within the gap an infinitely small amount of energy is sufficient for pair breaking, and thus dissipating quasiparticles would always be present even at very low temperatures.

4.2.2 Macroscopic Description

Characteristic quantities such as the penetration depth λ, coherence length ξ, critical current density j_c and critical magnetic field H_c can be well described by phenomenological macroscopic theories. These are the well-known London theory [174] and the Ginsburg–Landau theory [67]. The latter has been extended by Abrikosov [2] and Gorkov [73] and is therefore often called the GLAG theory. Abrikosov showed that there exists a periodic solution of the Ginsburg–Landau (G–L) equation which describes the Shubnikov phase. Gorkov proved that near to the phase transition the G–L theory is an exact approximation of the BCS theory. Further away from the phase transition this approximation is still valid within 25%. Both theories are three-dimensional isotropic continuum theories, which originally were not meant to describe anisotropic superconductors like the HTSCs [3].

One possibility for taking into account the anisotropy of the HTSCs is to replace the London penetration depth λ_L in the London equation

$$\Delta \times \boldsymbol{j} = \frac{1}{\mu_0 \lambda_L} \boldsymbol{B} \tag{4.1}$$

by different $\lambda_a, \lambda_b, \lambda_c$ for the different crystal axes. Thus initial qualitative statements can be made. For a more detailed consideration the distance s between the CuO_2 planes has to be compared with the coherence length ξ. For $YBa_2Cu_3O_{7-\delta}$ $\xi > s$ and thus the three-dimensional anisotropic G–L theory can be applied [39].

The isotropic G–L equations

$$\frac{1}{2m^*}\left|-i\hbar\nabla - e^*\boldsymbol{A}\right|^2 \psi + \alpha\psi + \beta|\psi|^2\psi = 0, \tag{4.2}$$

$$\boldsymbol{j} = -\frac{e^*i\hbar}{2m^*}(\psi^*\nabla\psi - \psi\nabla\psi^*) - \frac{e^{*2}}{m^*}|\psi|^2 \boldsymbol{A}, \tag{4.3}$$

with the boundary condition

$$\boldsymbol{n} = (-i\hbar\nabla - e^*\boldsymbol{A})\psi = 0 \tag{4.4}$$

and

$$\alpha = \frac{1}{\mu_0}\frac{H_{cth}^2}{n_S} \quad, \beta = \frac{1}{\mu_0}\frac{H_{cth}^2}{n_S^2}, \tag{4.5}$$

are modified by introducing the effective mass m^* as a tensor. Here n_S is the superconducting carrier density and e^* their charge, i.e. the charge of the Cooper pairs. It is assumed that this tensor can be normalized and thus $m_1 m_2 m_3 = 1$. This implies six *not* completely independent quantities

$\lambda_1, \lambda_2, \lambda_3$ and ξ_1, ξ_2, ξ_3. As $m_1 m_2 m_3 = 1$ it follows that $(\lambda_1 \lambda_2 \lambda_3)^{1/3} = \lambda$ and $(\xi_1 \xi_2 \xi_3)^{1/3} = \xi$. Moreover, for the Ginsburg–Landau parameter κ we obtain $\kappa = \lambda/\xi$. The thermodynamic critical field is then given by

$$H_{\text{cth}} = \frac{\Phi_0}{\sqrt{2}} 2\pi \lambda_i \xi_i \quad \text{for} \quad i = 1, 2, 3. \tag{4.6}$$

We now replace κ by $\kappa = \kappa_i \sqrt{m_i}$ in the isotropic G–L equation and thus obtain critical fields for each of the crystal axis directions $\hat{x}_i$ as follows:

$$H_{c1}(\| \hat{x}_i) = \frac{\Phi_0}{4\pi \lambda^2} (\ln \kappa_i + 0.5) \tag{4.7}$$

for the lower critical field, or, in terms of H_{cth},

$$H_{c1}(\| \hat{x}_i) = \frac{H_{\text{cth}}}{\kappa \sqrt{2}} (\ln \kappa_i + 0.5), \tag{4.8}$$

and for the upper critical field

$$H_{c2}(\| \hat{x}_i) = \frac{\Phi_0}{2\pi \xi_i^2}, \tag{4.9}$$

or, in terms of H_{cth},

$$H_{c2}(\| \hat{x}_i) = H_{\text{cth}} \kappa_i \sqrt{2}. \tag{4.10}$$

For fields with arbitrary orientations κ is replaced by $\bar{\kappa}(\Theta, \varphi)$, leading to the following expression:

$$\bar{\kappa}(\Theta, \varphi) = \frac{\kappa}{\sqrt{m_a \sin^2 \Theta \cos^2 \varphi + m_b \sin^2 \Theta \sin^2 \varphi + m_c \cos^2 \varphi}}. \tag{4.11}$$

This expression is in general not exact but provides a good approximation for superconductors with large κ at magnetic fields $H \gg H_{c1}$, and is exact at H_{c2}. Modifying the expressions for H_{c1} and H_{c2} accordingly leads to the following equation for H_{c2}, for example:

$$H_{c2}(\Theta, \varphi) = \sqrt{H_{c2a} \sin^2 \Theta \cos^2 \varphi + H_{c2b} \sin^2 \Theta \sin^2 \varphi + H_{c2c} \cos^2 \varphi}. \tag{4.12}$$

In the Shubnikov phase between H_{c1} and H_{c2} magnetic flux penetrates the superconductor as quantized flux lines. The core of these so-called Abrikosov vortices is normal-conducting as, according to the G–L theory, the superconducting order parameter goes to zero in the center of a flux line. If the superconductor is isotropic the cross section of a vortex is round. The anisotropic G–L theory implies an elliptical shape of the vortex cross section.

The anisotropic Ginsburg–Landau theory can no longer be applied if the distance s between copper oxide planes is greater than the coherence length

ξ, as is the case for the superconducting Bi, Tl and Hg compounds. In these compounds all relevant lengths are smaller than s and thus a basic assumption of the theory is not fulfilled. In these cases the *Lawrence–Doniach model* is applicable. The Lawrence–Doniach (L–D) theory [156] has been developed in order to describe layered superconductors. Superconductivity within planes is treated by the two-dimensional G–L theory and weak coupling is assumed between planes.

On the basis of this model, various authors have calculated the structure of flux lines in HTSCs ([18, 39] and references therein). A schematic view of this structure is given in Fig. 4.1. Within the ab planes the fluxons have the structure of pancake vortices. Between these planes they are weakly coupled by the Josephson effect or even only magnetically in the case of very high anisotropy. The interaction between pancake vortices in one plane is repulsive while it is attractive between pancakes in different planes. At $T=0$ and when the external magnetic field is oriented normal to the planes a stack of pancake vortices is obtained whose characteristics approximate these of an Abrikosov vortex in three dimensions (Fig. 4.1a). At elevated temperatures the pancake vortices become thermally activated and may shift with respect to each other (Fig. 4.1b). If the magnetic field is oriented parallel to the ab planes it is energetically favorable to arrange the flux lines between the planes. The flux lines no longer have a normal-conducting core as the order parameter does not become zero. They have a Josephson core and are therefore called Josephson

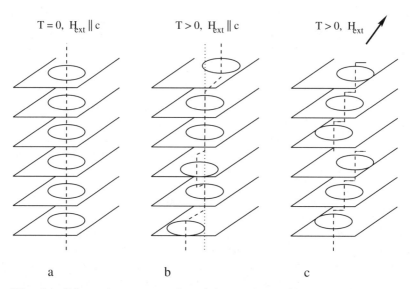

Fig. 4.1. Schematic representation of the structure of vortices in layered superconductors: **a)** at zero temperature in H_{ext} parallel to the c axis, **b)** at finite temperature in parallel magnetic field, and **c)** at finite temperature in arbitrary magnetic field

vortices. With arbitrary orientation of the external magnetic field the vortices form kinks (Fig. 4.1c). Pancake vortices within the planes are connected by Josephson vortices between the planes.

This special structure of vortices in HTSCs has consequences for application when dealing with flux dynamics and pinning properties. A more detailed discussion of these properties is hence necessary and will be given in the following section.

4.3 Electromagnetic Properties of HTSCs

4.3.1 Type II Superconductor in an External Magnetic Field

The properties of type II superconductors relevant for technical applications are their behavior in a magnetic field and when carrying a transport current. The two principal properties of a superconductor are the capability of carrying currents without losses and expelling a magnetic field from its interior (Meissner–Ochsenfeld effect [192]). Two parameters decide between two totally different states of a superconductor in an external magnetic field: the Ginsburg–Landau parameter κ, which was introduced in Sect. 4.2.2, and the energy of the phase boundary between the normal and superconducting phases

$$\sigma_{\mathrm{ns}} = (\xi - \lambda)\frac{1}{2\mu_0}H_{\mathrm{cth}}^2. \qquad (4.13)$$

If the coherence length $\xi > 1$ and thus the phase boundary energy σ_{ns} positive, the superconductor is of type I. In an external magnetic field it exhibits the Meissner–Ochsenfeld effect up to the upper critical field H_{c}. If this critical field is exceeded locally owing to demagnetization effects the superconductor goes into the *intermediate state*. The characteristics of the intermediate state is that, owing to the positive value of the phase boundary energy, the normal and superconducting states coexist on a macroscopic scale. A more detailed discussion of the intermediate state and the calculation of the energy σ_{ns} is given by Buckel [24] and Abrikosov [3].

For $\kappa > 1/\sqrt{2}$ and $\sigma_{\mathrm{ns}} < 0$ we are dealing with a type II superconductor. The consequence of the negative phase boundary energy is that it becomes energetically favorable to let magnetic flux into the interior of the superconductor as soon as the lower critical field H_{c1} is exceeded. This happens in microscopic quantities of size $\Phi_0 = h/2e$, called flux quanta or Abrikosov vortices. The superconductor goes over to the mixed state or *Shubnikov phase*. The vortices arrange themselves in a triangular flux line lattice (FLL). Abrikosov [2] showed that this state is an exact solution of the G–L equations (4.2), (4.3) for $\kappa > 1/\sqrt{2}$. When the external field is increased, more and more flux is introduced into the superconductor up to the upper critical field H_{c2}.

There the fluxoids start to overlap and superconductivity is destroyed. The magnetic phase diagram of a type II superconductor is depicted in Fig. 4.2. Below H_{c1} the superconductor is in the Meissner state, between H_{c1} and H_{c2} it forms the Shubnikov phase, becoming normal-conducting only above H_{c2}. The triangular lattice of the flux lines could be demonstrated in many experiments, for example by Essmann and Träuble [48, 49].

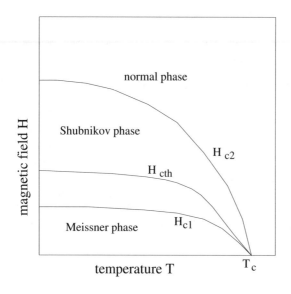

Fig. 4.2. Schematic phase diagram of a type II superconductor

Only type II superconductors are relevant for applications in magnet and energy technology as the critical fields of type I superconductors are very low (e.g. Pb: 80 mT at $T = 0$). The production of high-field magnets is not possible with type I materials. The critical fields H_{c1} and H_{c2} can be derived from simple assumptions. As soon as one fluxon penetrates the superconductor H_{c1} is reached. Calculating the magnetization of this single vortex leads to [3]

$$H_{c1} \approx \frac{H_{cth}}{\kappa}(\ln \kappa + 0.5) \,. \tag{4.14}$$

Under the influence of an external magnetic field, Cooper pairs move in a circle. The radius of their orbit depends on the magnetic field strength. As soon as the radius of this orbit becomes smaller than the spatial extension ξ of the Cooper pair or if its kinetic energy becomes larger than the gap energy Δ, the pairs break up and the superconductor becomes normal. This is the criterion for estimating H_{c2} [3]:

$$H_{c2} \approx H_{cth}\kappa \,. \tag{4.15}$$

In general it can be said that the greater κ is, the smaller H_{c1} and the larger H_{c2}. The HTSC materials have very high κ values of around 1000 and therefore their lower critical fields are very small ($\mu_0 H_{c1} \approx 100\,\mathrm{mT}$ for $YBa_2Cu_3O_{7-\delta}$) and their upper critical fields are very high (around 50 T for $YBa_2Cu_3O_{7-\delta}$ and nearly 100 T, presumably, for $(Pb,Bi)_2Sr_2Ca_2Cu_3O_{10}$).

When the superconductor is carrying a current I the flux lines experience a Lorentz force

$$F_\mathrm{L} = \boldsymbol{I} \times \boldsymbol{B} \tag{4.16}$$

with $B = \mu_0 H$. Under this force the flux lines start to move through the lattice and as their core is normal-conducting they dissipate energy. This means that as soon as vortices move through the superconductor the losses are no longer zero. In an ideal type II superconductor there is nothing to hinder the motion of flux lines; instead they can move freely, which is equivalent to a vanishing critical current I_c. Such superconductors are not relevant for applications.

In a real type II superconductor, as in any other solid, there are always lattice distortions and impurities with reduced superconducting or even normal-conducting properties. The superconducting order parameter is either reduced or zero, as within a vortex core. That implies that such defects are energetically favorable places for vortices to reside and the vortices will be pinned in the potential of these so-called pinning centers. The efficiency of such a pinning center is at its maximum if its size is of the order of the superconducting coherence length ξ. In conventional superconductors ξ is around 50–500 nm. Thus precipitates are ideal pinning centers. In HTSCs with $\xi \sim 0.3$–3 nm, pinning sites have to be of atomic order, such as dislocations, stacking faults, twin boundaries or vacancies. Such sites are much harder to produce. If pinning is efficient the critical current density j_c becomes high and the material is interesting for applications. Type II superconductors with good pinning properties, i.e. a high critical current density j_c, are called *hard superconductors*.

On one hand the properties of the flux line lattice and the pinning properties are important for applications; on the other hand they are complex and interesting topics of condensed-matter physics and materials science. Hence these subjects will be treated in more detail in the following sections.

4.3.2 Elastic Properties of the Flux Line Lattice

Analogously to a crystal lattice, the elastic energy of a flux line lattice may be calculated using linear elasticity theory [22]. Pinning, thermal activation and structural defects introduce small perturbations. The elastic energy in k space is deduced from the Fourier components of the vortex displacement. In the case of a uniaxial elastic medium we obtain the following elastic matrix:

$$\Phi_{xx} = c_{11}k_x^2 + c_{66}k_y^2 + c_{44}k_z^2 + \alpha_\mathrm{L}(k)\,, \tag{4.17a}$$

$$\Phi_{yy} = c_{66}k_x^2 + c_{11}k_y^2 + c_{44}k_z^2 + \alpha_\mathrm{L}(k)\,, \tag{4.17b}$$

$$\Phi_{xy} = \Phi_{yx} = (c_{11} - c_{66})k_x k_y\,. \tag{4.17c}$$

For completeness the *Labusch parameter* α_L [151] has been added, which describes the interaction of the FLL with a pinning center induced by material inhomogeneities. The quantities c_{11}, c_{44} and c_{66} denote the compressional, tilt and shear moduli, respectively. For the tilt and compressional moduli

$$c_{11} \approx c_{44} \approx \frac{B^2}{\mu_0} \tag{4.18}$$

is valid only in the isotropic case in the small region in the middle of the Brillouin zone (BZ). On the edge of the BZ the boundary condition of periodicity implies deviation from quadratic behavior. The moduli become dispersive. The physical reason for this behavior is that the characteristic interaction length of electromagnetic fields with the vortex lattice (*the Campbell penetration depth* λ_C [29, 30, 31]) is large compared with the mean vortex spacing $a_0 = 1/\sqrt{3}B^2$. That means that the FLL becomes soft for short-wavelength excitations.

The general G–L solution for the compressional and tilt moduli is

$$c_{11} \approx c_{44} \approx \frac{B^2}{\mu_0} \sum_{n=1}^{\infty} (-1)^{n-1} \left(\frac{|\psi|^2}{k^2 \lambda^2} \right)^n. \tag{4.19}$$

If k or λ goes to infinity, which is equivalent to B approaching $\mu_0 H_{c1}$ or $\mu_0 H_{c2}$, respectively, the elastic moduli approach zero. So for long-wavelength excitations the FLL becomes soft, too.

In the limit of vanishing induction B the line tension of a single vortex is $P = c_{44}\Phi_0/\mu_0 H_{c1}$. In anisotropic superconductors P is small compared with the self-energy $\epsilon_S = \Phi_0 \mu_0 H_{c1}$. The self-energy ϵ_S also becomes dependent on angle for a stack of pancake vortices. Consequently the line tension and tilt modulus are strongly reduced in anisotropic materials. So in layered superconductors it becomes very easy to remove a pancake vortex from the stack.

In the case of single-vortex pinning, pinning sites can be described as parabolic potentials in which the flux line is bound. If direct summation of of pinning forces is possible the Labusch parameter α is a measure of the slope of the pinning potentials. The problem of summation of the energy contributions of the single pinning potentials to an overall pinning force was discussed in some detail by Larkin and Ovchinnikov [155].

4.3.3 Phase Transitions in the Vortex Lattice

At high temperatures, owing to the high anisotropy and soft modes, fluctuations become large in the FLL of HTSCs. In order to estimate the fluctuations the mean square displacement of the vortices $\langle u^2 \rangle$ of each elastic mode of the lattice can be associated with the energy $k_B T/2$. The result is expressed by the *Ginsburg number*

$$Gi = \left(\frac{k_{\mathrm{B}}T}{2\mu_0 H_{c2}\, \xi_{ab}\, \xi_c} \right)^2, \tag{4.20}$$

which denotes the relative width of the fluctuation region near to H_{c2}. For comparison, in conventional superconductors Gi is of the order of 10^{-8}, while in high-temperature superconductors it is of order 10^{-2}.

Thus the specific properties of the HTSCs such as high transition temperature and short coherence length also lead to large thermal fluctuations. The timescale of these fluctuations is determined by the elastic restoring force and the viscous motion of the vortices within the FLL. A flux line moving with velocity v exerts a force ηv on the crystal lattice. This leads to melting of the FLL appreciably below T_c. This melting transition is a first-order phase transition in pure materials; in the presence of pinning and strong disorder it turns into a second-order phase transition. A schematic outline of the vortex phase diagram in an HTSC is given in Fig. 4.3.

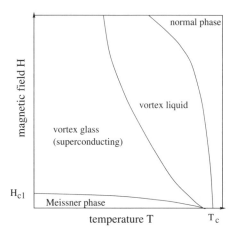

Fig. 4.3. Hypothetical vortex phase diagram in an HTSC

In a pure superconductor the first-order phase transition from vortex solid to vortex liquid is expected to obey the *Lindemann criterion* [22, 171]. This criterion says that a lattice will melt as soon as the ratio between kinetic and potential energy exceeds a certain value c_{L}, or, more physically, if the lattice vibrations or thermal fluctuations are of the order of the binding energy. The melting temperature T_{m} follows from the mean square displacement $\langle u^2 \rangle$ and the mean vortex spacing a_0. Thus the Lindemann criterion is

$$\langle u^2 \rangle^{1/2} = c_{\mathrm{L}} a_0 . \tag{4.21}$$

The first thermodynamic evidence for this first-order phase transition in Bi-2212 single crystals was found in experiments by Zeldov and coworkers [282].

As a consequence of the layered structure of the HTSCs there is also a decoupling transition at which the pancake vortices of adjacent CuO_2 planes decouple and the vortex characteristics change from three-dimensional to two-dimensional.

4.4 Flux Line Dynamics

Up to now only reversible processes in the FLL have been discussed. If a current is applied to a hard type II superconductor which is higher than its critical current, irreversible processes occur. As mentioned above, the current induces a Lorentz force on the flux lines (4.16). Often the Lorentz force density,

$$\boldsymbol{f}_\mathrm{L} = \boldsymbol{j} \times \boldsymbol{B}\,, \tag{4.22}$$

is considered instead. Owing to the current density j, vortices can escape from their pinning potential and move through the superconductor with a velocity v. This motion generates an electric field

$$\boldsymbol{E} = \boldsymbol{v} \times \boldsymbol{B}\,. \tag{4.23}$$

Moving flux lines dissipate energy because of their normal-conducting core, as Bardeen and Stephen [13] showed.

In this regime of flux flow, vortices move in bundles. Even avalanche effects can happen, as during magnetizing of large YBCO bulk samples [268]. In these experiments the avalanches were so great that the sample was destroyed by the dissipated energy.

A fundamental treatment of the dynamics of vortices in hard superconductors was provided by the basic models of Bean [14] and Anderson and Kim [10]. These models are still valid today and will therefore be discussed in some detail.

4.4.1 Flux Pinning and the Bean Critical-State Model

Bean proposed a model to describe the magnetization of hard superconductors in alternating magnetic fields. Typical magnetization curves of type I and type II superconductors are depicted in Fig. 4.4. The *Bean critical-state model* serves as a straightforward phenomenological explanation for many experimental results on hard superconductors in the critical state (Shubnikov phase), even for many results on HTSC materials. It does not, however, give any explanation for the microscopic origin of pinning.

These are the basic assumptions of the Bean critical-state model:

- Only the critical state is considered. There is no reversible magnetization; $H_{c1} = 0$.
- There is a critical current density which can be carried by the superconductor.

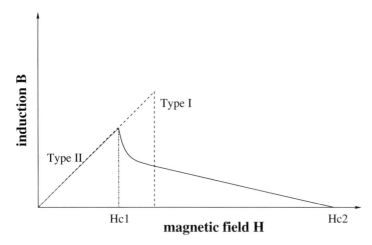

Fig. 4.4. Typical magnetization curves for type I and type II superconductors in an external magnetic field (schematic)

- Even very low electromagnetic forces induce the full critical current density.
- The critical current density is independent of the external magnetic field.
- In any region of the superconductor where magnetic flux has penetrated, the critical current density is perpendicular to the magnetic induction B; everywhere else $j_c = 0$ is valid.

On the basis of these assumptions, the magnetization in a parallel magnetic field is calculated for a disk of thickness d or a cylinder of radius r. These geometries are chosen because of their very low demagnetization effect. Figure 4.5 shows the flux profile and critical current density distribution in a cylinder of radius r in an alternating external magnetic field H_{ext}. The following can be seen in Fig. 4.5 from left to right:

1. With increasing external field H_{ext}, magnetic flux penetrates the sample and hence the remanent flux or induction B decreases linearly towards its center. The slope of the induction corresponds to the critical current density j_c in the region where flux is already present.
2. As soon as the external field reaches the value H^* the cylinder is totally penetrated by flux. This field is called the full-penetration field.
3. Further enhancement of H_{ext} leads to the introduction of more flux into the sample.
4. When the external field decreases the flux profile is inverted and the currents change their direction.
5. At $H_{\text{ext}} = 0$ there is still flux inside the superconductor and the flux distribution takes the form of a cone, the so-called *Bean cone*.

Calculation of the magnetization of a hard superconductor from a series of flux distributions and critical current density profiles yields a diamagnetic

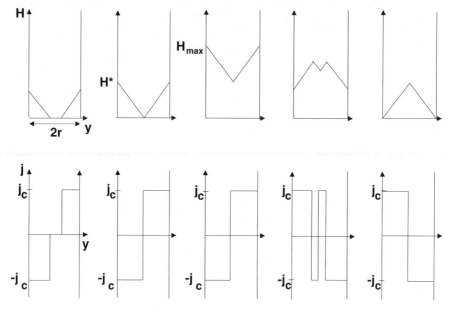

Fig. 4.5. Magnetic flux profile and critical current density distribution of a superconducting cylinder in an alternating external magnetic field H_{ext}

hysteresis analogous to the ferromagnetic hysteresis of magnets. The area of the hysteresis is a measure of the energy dissipated owing to flux motion:

$$W_{mag} = \oint H dB, \tag{4.24}$$

$$W_{mag} = \frac{H_{ext}^3}{j_c}. \tag{4.25}$$

This enables the quantitative determination of j_c from magnetization hysteresis, but only where the magnetization curve does not have a significant slope (see Fig. 4.4):

$$j_c = \frac{\Delta M}{2r}. \tag{4.26}$$

Nevertheless, there are some important limitations to the Bean critical state model:

- There is no Meissner state in Bean's considerations.
- In HTSCs the critical current density depends strongly on the external magnetic field.
- The model does not consider the origin of the critical current density.

These shortcomings are partly overcome by the model proposed by Anderson and Kim [10], which will be described in the next section.

4.4.2 Thermally Activated Flux Creep

Anderson [9] and Anderson and Kim [10] developed a theory of vortex motion in hard superconductors on the basis of the GLAG theory. This theory of thermally activated flux creep, though originally meant for conventional superconductors, nevertheless still applies well for the HTSCs. Many theories of vortex dynamics in high-temperature superconductors are based on this *Kim–Anderson model*. A very extensive review on vortices in HTSCs is given by Blatter et al. [18]. The present work, however, will focus on the treatment of the Kim–Anderson model and some special features of HTSCs. The principal goal of the model is to calculate creep rates of magnetic flux or flux bundles for different limits in an existing Abrikosov lattice. The following assumptions are made:

- A bundle of flux lines of size $\lambda > \xi$ is bound to a pinning center. A smaller flux bundle size is not probable as distortions of the FLL lower in size than the coherence length would need too much energy for their formation.
- The overall pinning potential is periodic and either harmonic or of zigzag form.
- The Lorentz force acts on the flux bundles.

Figure 4.6 shows a periodic harmonic pinning potential with different applied current densities. This figure illustrates that the Lorentz force changes the effective barrier height between neighboring pinning potentials, which may be of the order of ξ with p being the number of active pinning sites:

$$U_0 = \frac{1}{2} p H^2 \xi_0^3 . \tag{4.27}$$

Under the condition that j_c and the external magnetic field H_{ext} are normal to each other, flux bundles having a size λ and length l experience the following Lorentz force:

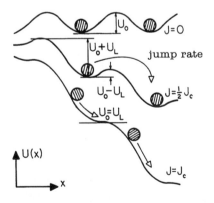

Fig. 4.6. Periodic pinning potential under the influence of different current densities. Jump rates are calculated from the effective barrier heights (after [173])

$$F_L = j H_{ext} \lambda^2 l . \tag{4.28}$$

The effective barrier height is then determined from the difference between the barrier height and the energy due to the Lorentz force:

$$U_{eff} = U_0 - U_L = \frac{1}{2} p H_{ext}^2 \xi_0 - j H_{ext} \lambda^2 \xi_0 . \tag{4.29}$$

At nonzero temperature some flux bundles are activated and are able to move through the FLL. This thermally activated process obeys a rate equation:

$$R = \omega_0 \, exp \frac{-U_0}{k_B T} . \tag{4.30}$$

This is the jump rate R of the vortex bundles, with the vibration frequency $\omega_0 \approx 10^5 - 10^{10}$ Hz in LTSC. The motion of flux bundles dissipates energy. This regime is called the *flux creep* regime.

If the critical current density j_c is reached, the Lorentz force density is balanced by the pinning-force density

$$\boldsymbol{f}_L = \boldsymbol{f}_p = \boldsymbol{j}_c \times \boldsymbol{B} , \tag{4.31}$$

and thus $U_L = U_0$ and vortices can move freely through the superconductor. This state is called *flux flow* and appears shortly before the material becomes normal-conducting. In conventional superconductors the flux flow can only be observed in a very small regime below the superconducting–normal transition. In HTSCs, however, the flux flow regime can occupy a rather large field and temperature range (*giant flux creep*).

Thermally activated flux creep causes an electric field. The Lorentz force is needed to calculate this field from the creep rate. The jump rate in the direction of the Lorentz force is

$$R_1 = \omega_0 \exp \frac{-(U_0 - U_L)}{k_B T} \tag{4.32}$$

and against the force direction

$$R_2 = \omega_0 \exp \frac{-(U_0 + U_L)}{k_B T} . \tag{4.33}$$

The field E is then determined from the effective creep rate $R_{eff} = R_1 - R_2$ and the drift velocity $v = R_{eff} l$, using $E = vB = \rho j$, as

$$E(j) = 2 \rho_c j_c \exp\left(\frac{-U_0}{k_B T}\right) \sinh\left(\frac{jU_0}{j_c k_B T}\right) . \tag{4.34}$$

The critical current density j_0 at $T=0$, the resistance ρ_c at j_c and the pinning energy U_0 are phenomenological parameters. Equation (4.34) is the basic equation of the Kim–Anderson model, from which all special cases are derived.

Three important limiting cases have been formulated by Brandt [21]. For small current densities $j \ll j_c$, the sinh function can be linearized. We obtain ohmic behavior with a thermally activated linear resistivity ρ_{TAFF}. For large current densities of the order of j_c the equation can be approximated by $E \propto \exp(jU_0/j_c k_B T)$, expressing the sinh as an exponential function. This is the flux creep regime. For very large current densities $j \gg j_c$ we obtain the usual flux flow resistivity ρ_{FF}. Summing up, we obtain the following equations for the different regimes of flux line dynamics:

$$\rho = \frac{2\rho_c U_0}{k_B T} \exp\left(\frac{-U_0}{k_B T}\right) = \rho_{\mathrm{TAFF}} \tag{4.35}$$

$$\text{for} \quad j \gg \frac{j_c k_B T}{U} = j_1 \quad (\text{TAFF}),$$

$$\rho = \rho_c \exp\left[\left(\frac{j}{j_c} - 1\right)\frac{U}{k_B T}\right] \propto \exp\left(\frac{j}{j_1}\right) \tag{4.36}$$

$$\text{for} \quad j \approx j_c \quad (\text{flux creep}),$$

$$\rho = \rho_{\mathrm{FF}}\left(1 - \frac{j_c^2}{j^2}\right)^{1/2} \approx \rho_{\mathrm{FF}} \approx \rho_n \frac{B}{B_{c2}(T)} \tag{4.37}$$

$$\text{for} \quad j \gg j_c \quad (\text{flux flow}).$$

The E–j characteristics of the different regimes are depicted in Fig. 4.7.

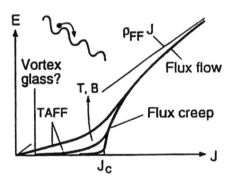

Fig. 4.7. E–j characteristics for the different regimes of flux motion (after [21])

More recent models specifically developed to explain the unique properties of the HTSCs, such as the collective-creep model and the vortex glass model, go beyond the Kim–Anderson model. Nevertheless similar E–j characteristics are obtained in both descriptions. The physical picture behind the collective-creep model [53] is that weak random pinning acts on the FLL. Thus the FLL can be treated as an ideal elastic medium, which also can be isotropic. In this picture the thermally activated resistivity really approaches zero. The vortex glass picture gives at least qualitatively a similar behavior. This model is based on the assumption of a second-order phase transition in the FLL from a vortex glass to a vortex liquid. This theory was derived by Fisher, Fisher and Huse [54] in analogy to the theory of spin glasses.

The basic assumptions of both theories are the following. Characteristic quantities such as the pinning correlation length ξ_g (size of the jump volume) and the relaxation time τ_g of the fluctuations of the glassy order parameter diverge near to the glass temperature T_g:

$$\xi_g(T, B) \approx \xi_g(B) \left|1 - \frac{T}{T_g}\right|^\nu, \tag{4.38}$$

$$\tau_g(T, B) \approx \tau_g(B) \left|1 - \frac{T}{T_g}\right|^{\nu z}. \tag{4.39}$$

A scaling behavior of the E–j curves is predicted. The electric field should scale like $E \xi_g^{z-1} = f_\pm (j \xi_g^{D-1})$, with $z \approx 4$ and D the spatial dimension. The functions f_+ and f_- are the two master curves for the regimes above and below T_g. For $x \to 0$, $f_+(x) =$ const. and $f_-(x) \to \exp(-x^{-\nu})$. At $T = T_g$, E obeys a power law: $E \propto j^{(x+1)/(D-1)}$. Thus the resistivities turn out to be

$$\rho \propto j^{(x+1)/(D-1)-1} \quad \text{for} \quad T = T_g, \tag{4.40}$$

$$\rho \propto \exp[-(j_c/j)^\alpha] \quad \text{for} \quad T < T_g. \tag{4.41}$$

There is no explicit glass temperature within the framework of the collective-creep model. The picture, however, is similar as flux creep occurs only above a certain melting temperature T_m, above which the FLL loses its elastic rigidity. This model will be used to discuss experimental E–j characteristics in a later chapter.

Part II

Recent Achievements

5. Conductor Preparation and Phase Evolution

The development of materials processing and characterization in the field of HTSCs for magnet and energy technology has been fairly fast in recent years. As the production of long lengths of conductors and of large monolithic ceramics is also of commercial interest a lot of the present know-how belongs to companies and is thus often confidential. Consequently, the overview of recent results given below, based on published literature, may not be complete concerning unpublished knowledge within companies. In addition, owing to the vast amount of literature, this review concentrates on the information considered to be the most important for recent developments.

5.1 Preparation of Wires, Tapes and Bulk HTSCs

Prior to application of any kind is the preparation of suitable material from which systems can be developed. Consequently the focus has been on the development of conductor materials and bulk ceramics of HTSCs with high critical current densities and strong flux-pinning characteristics. As already discussed, the preparation routes are very different for Bi-2223, Bi-2212 and YBCO. The state of the art will be discussed below in separate sections.

5.1.1 Preparation of Bi-2223 Tapes

By far the most effort has been put into the development of tape material of the lead-doped Bi-2223 superconductor as this material has the highest transition temperature (110 K) of the technically relevant HTS compounds. Owing to the high transition temperature and the prospect of high critical current densities the Bi-2223 tapes are expected to find their main application in the field of energy technology. The preparation procedure includes several steps: the raw materials such as precursor powders and sheath materials, filling and mechanical deformation and, finally, the thermo-mechanical treatment.

Precursors

Properties of the starting powder such as stoichiometry, phase purity and particle size strongly influence the phase development of the superconductor.

There are several important points. Firstly, the powder may not be completely reacted Bi-2223; this is disadvantageous for the quality of the final conductor. Precursor powders therefore are compositions of Bi-2212 and other components, which will then be reacted to Bi-2223. Secondly, lead has to be added to the overall composition in order to stabilize the Bi-2223 phase as described earlier. Thirdly, the particle size of the powders is important. Usually a grain size of 1–5 µm is used.

The typical nominal composition commonly used [55, 234, 280] is $Bi_{1.8}Pb_{0.34}Sr_{1.9}Ca_2Cu_{3.1}O_5$ or similar. This composition is chosen in order to satisfy the following relations: the content of Bi and Pb together has to exceed 2.0 as some lead and bismuth are lost during processing; the content of Ca must exceed the Sr content in order to increase the slow Ca diffusion; and the copper content has to be more than 3.0 because the copper surplus is needed to form the liquid phase which supports the Bi-2223 phase formation during the initial stages.

Moreover, not only the stoichiometry but also the phase composition of the starting powder is of great importance. Starting with fully reacted Bi-2223 does not result in high critical current densities. There are two possibilities for the starting compositon, one where the lead is incorporated into the Bi-2212 phase and one where the lead is present as an alkaline-earth compound. These compositions are:

- Bi(Pb)-2212 + Ca_2CuO_3 + CuO
- Bi-2212 + Ca_2PbO_4 + CuO.

It has been shown by Dorris et al. [43] that the reaction can be faster if the lead is already incorporated into the Bi-2212, as the first stage of the Bi-2223 reaction is the incorporation of lead as shown by Jeremie et al. [110].

Sheath Material

Usually pure silver is preferred as a sheath material as it does not show any reaction with the ceramic core. However, there are some difficulties with this material. First, for considering cryogenic applications like current leads, the high thermal conductivity of silver is a disadvantage. For this purpose Ag/Au alloys with 11% Au are used by most groups [185]. A further disadvantage is the softness of silver. In particular, for applications in magnet and energy technology the tape material has to withstand mechanical stress if reliable power cables or magnets are to be built. Several alloys have been tested by many groups, but at present an Ag/Mg alloy with less than 2% Mg is used [69, 229]. During sintering the magnesium reacts to form MgO, which leads to precipitation hardening. The drawback of this alloy is that there are reactions between the Mg and the superconductor. To circumvent this problem only the outer sheath is alloyed, while the sheaths of the individual filaments are pure silver [229].

5.1 Preparation of Wires, Tapes and Bulk HTSCs

A third important problem arises from the interfilamentary coupling. Owing to the softness of the silver sheath, bridging of needle-like Bi-2223 crystallites from one filament to a neighboring one is observed fairly often. This microstructural feature leads to severe ac losses. Huang and Flükiger [104, 103] succeeded in covering the individual filaments with a resistive barrier of $BaZrO_3$, which prevents bridging and significantly reduces ac losses. This method is now also adopted by other groups [138].

Mechanical Deformation

The mechanical-deformation process plays an important role in the fabrication of long lengths of BSCCO tapes and wires. The cross sections and longitudinal sections of the tapes and wires have to be very homogeneous in order to provide the largest possible area for current transport. Furthermore, the alignment and texture of the BSCCO crystallites have to be as good as possible to achieve good grain contacts. Sausaging, pure texture and thickness inhomogeneities of the filaments are the main sources of low critical current densities. Very good results in the rolling of long tapes of Bi-2223 conductors are achieved by companies such as American Superconductor Corp. (ASC) and NST. But only NST has published a systematic study and the results of the deformation process [84]. The NST researchers calculated and measured the stress and strain distribution during roling and pressing and from this established three important observations, which also had been made earlier [215]:

- The stress distribution during pressing is such that microcracks are preferentially induced along the tape and thus do not hinder the current path significantly.
- The stress distribution during rolling is such that microcracks are introduced normal to the current path, which obviously is deleterious.
- Over the cross section of a tape the rolling or pressing force is highest in the middle, consequently leading to less deformed filaments at the tape edges.

Thermo-Mechanical Treatment

The world best Bi-2223 tapes with the highest j_c are now made by ASC, closely followed by NST. A few years ago the highest value for the critical current density was reported by Li et al. [165] to be $69\,000\,A/cm^2$ at 77 K in self-field in short pressed samples. ASC has exceeded this value only recently, by achieving $70\,000\,A/cm^2$ in short rolled samples [234]. The highest j_c values of $24\,000\,A/cm^2$ in long tapes have been obtained by NST [83]. Besides the deformation process, thermo-mechanical treatment also plays an important role in achieving these results. In recent years a certain standard processing scheme has been developed. The processing atmosphere is either air, or argon with 8% oxygen, or even an atmosphere with variable oxygen content. The processing temperature is around 845°C in air and around 830°C in a reduced-oxygen atmosphere. The tapes are sintered for 60 to 80 hours depending on the phase composition of the starting powders. An intermediate rolling step is applied after 30 to 50 hours of sintering time. Companies that fabricate long lengths of tape material usually do their processing in a reduced-oxygen atmosphere (e.g. [52]) as in this case the temperature window where the phase formation of Bi-2223 takes place is much broader, namely $\Delta T = 15°C$ in reduced oxygen as compared with 5°C in air. Consequently, in a reduced-oxygen atmosphere, processing is less critical.

Recently, a so-called post-annealing process has been introduced. Wang et al. [265] have shown that adding a processing step at a lower temperature in the final heat treatment enhances the critical current density by reducing second-phase precipitates of Bi-2201 at grain boundaries. A real post-annealing at about 820°C in air after the final sintering step has been shown to enhance j_c by a factor of three [159]. In this case not only Bi-2201 but also Bi-3221 was reduced at the grain boundaries. These results will be presented in Chap. 9.

5.1.2 Preparation of Bi-2212 Tapes and Wires

The preparation of Bi-2212 follows a somewhat simpler processing scheme than the Bi-2223 processing. The starting powder is stoichiometric or sometimes slightly off-stoichiometric fully reacted Bi-2212. For the powder particle size distribution the same arguments hold as for Bi-2223, and for the deformation process the same considerations as above are valid. Up to the thermal treatment there is no significant difference in the processing of the two BSCCO materials if the PIT process is applied. However, dip coating is a real alternative processing route for Bi-2212 tapes [86].

The Bi-2212 tapes with the world highest critical current density are produced by dip coating with a preannealing and intermediate rolling process (PAIR), introduced by Kitaguchi et al. [122, 123, 124, 193, 194, 195] of Furukawa Electric in Japan. Thin silver ribbons are dip coated with reacted Bi-2212 ink and sandwiched to form multifilamentary tapes. These tapes are then preannealed at 700°C in vacuum and at 835°C in pure oxygen. After this preannealing an intermediate rolling step is applied. The tapes are subsequently melt processed in pure oxygen. With this procedure the authors achieve j_c values of up to $900\,000\,\text{A/cm}^2$ at 4.2 K in self-field and $500\,000\,\text{A/cm}^2$ in 10 T. As dip coating is a process which can easily be scaled up to produce long lengths of tape this process is highly interesting for the production of high-field magnets.

The highest critical current densities for PIT tapes and wires have been obtained by Okada and coworkers [205] of Hitachi, also in Japan. With a standard process of preannealing in vacuum and pure oxygen with subsequent melt processing in pure oxygen they obtain critical current densities in the range of 450 000 A/cm^2 at 4.2 K in self-field. The first test magnets, as insert coils in very-high-field magnets, have been built with this material in cooperation with NRIM (National Research Institute of Metals) in Japan. Fields of up to 22 T have been obtained [143].

5.1.3 YBCO-Coated-Conductor Preparation

Because of its better flux-pinning behavior YBCO has high critical current densities in higher magnetic fields and at higher temperatures than the BSCCO conductors. However, its grain boundaries behave much less simply. Even slight misorientations completely interrupt the supercurrent, as discussed earlier. Therefore the PIT technique is not a good method to produce YBCO conductors. A thin- or thick-film technique has to be used instead. Initial promising results have been obtained with a method first introduced by Iijima et al. [108], which they called IBAD (ion-beam-assisted deposition). With this technique, epitaxial YBCO films are grown on top of polycrystalline metallic substrates such as hastalloy or nickel covered with a buffer multi-layer of YSZ–CeO$_2$–YSZ. The insulating buffer layers are epitaxially grown by IBAD. On top of well-aligned buffer layers, YBCO deposited by laser ablation or pulsed laser deposition (PLD) also develops a well-defined texture and thus high critical current densities of up to 2.8 MA/cm^2 are reached [57, 212]. At present it is possible to deposit high-j_c YBCO films on large-area technical substrates [58] and on up to one meter of flexible metal substrates [57]. Despite the fact that the current densities which can be achieved by this method are very high, there is a big disadvantage. IBAD is a very slow process. It takes about one hour to fabricate one centimeter of a flexible YBCO tape.

In 1996 a group at Oak Ridge National Laboratory succeeded in depositing high-critical-current-density YBCO films on biaxially aligned substrates consisting of recrystallized nickel tapes with a pronounced cube texture and a buffer multilayer of CeO$_2$ and YSZ [76]. They called this process RABiTS^{TM}. Recrystallization texturing of fcc metals has been well known since the early 1920s and 1930s [70, 71, 72, 200]. The pronounced cube texture of the nickel tapes makes it possible to grow well-aligned buffer layers with conventional sputtering or evaporation techniques. The YBCO films deposited on top of these RABiTS^{TM} tapes also exhibit good texture. This process is much faster than the IBAD process. However, up to now no tapes longer than ten centimeters have been fabricated which still have good current-carrying capabilities. It has to be mentioned, though, that owing to their superior in-field properties the YBCO tapes will be a much better candidate for high-magnetic-field applications, especially at liquid-nitrogen temperature, than the BSCCO conductors. But up to now long lengths of flexible and technically useful tape conductors have only been fabricated out of Bi-2223.

76   5. Conductor Preparation and Phase Evolution

5.1.4 Melt Processing of Bulk REBCO

As described in the previous section, the preparation of $YBa_2Cu_3O_{7-\delta}$ is much easier than that of BSSCO. Consequently, the fabrication of large amounts of bulk material of this compound has made great progress in recent years. Large monolithic monodomain samples have been prepared by several groups in the world [172, 203, 235]. The YBCO ceramics usually have Y-211 inclusions and additions such as CeO_2 or PtO_2. This phase composition ensures a fine, submicron-sized distribution of the Y-211 inclusions which serve as pinning centers. These samples have high critical current densities and show good levitation force properties.

In addition, Nd-123 has turned out to be a compound with even better pinning properties, especially at elevated magnetic fields. In the beginning the phase-pure preparation of this compound was difficult because the ionic radii of Nd and Ba are similar, which leads to the incorporation of Nd on Ba sites. Now that this problem has been overcome [254, 279], this material seems to be even more suitable for applications than Y-123.

Recently Murakami reported very promising results on RE-123 compounds with up to three rare-earth elements incorporated into the crystal lattice. These materials seem to have superior superconducting and mechanical properties [203, 240].

5.2 Phase Formation and Microstructure

The first success in improving the material properties of the ceramic superconductors was more or less empirical. Changing processing parameters and measuring their influence on the critical current density lead to a fairly fast improvement of the material. However, since the results of Li et al. in 1993 [165], the highest j_c achieved in Bi-2223 tapes is around $70\,000\,A/cm^2$ at 77 K in self-field. The only progress which has been made in the meantime is that this value, first obtained in short pressed samples, now also can be achieved in short rolled samples [234]. The reason for this is that a detailed understanding of the phase formation of Bi-2223 in the PIT process is still missing. In addition, the microstructure which enables efficient transport of supercurrent is also not very well understood. Further improvement of the critical current density seems to be possible only with a clear understanding of the material properties of the BSCCO conductors. For this reason several groups have started to look more closely at the phase formation and microstructure, of Bi-2223 in particular, in order to find the current-limiting mechanisms.

5.2.1 Observation of Phase Evolution of Bi-2223

A lot of important work on the phase formation of Bi-2223 has been performed by the Geneva group [55]. By combined analytical methods, e.g.

electron microscopy and x-ray diffraction of pressed powder samples and Bi-2223 tapes, they determined the phase evolution of Bi-2223 in a silver sheath as described in Sect. 7.1. There has also been work by other authors [46, 77, 266], some of them proposing different formation mechanisms for the Bi-2223 phase. Especially under discussion is the question whether the Bi-2223 phase is formed by nucleation and growth or by intercalation of one additional $(Sr,Ca)CuO_2$ layer into Bi-2212 grains [28]. The latter process should be very slow. At present it seems to be that both formation mechanisms are possible, depending on the phase composition of the starting powder. If one starts with plumbate phases the reaction is supported by transient liquid and is faster but the product may contain more second phases. If the lead is incorporated into the Bi-2212 phase and the oxygen stoichiometry is chosen to be ideal, the reaction is slower as it is no longer liquid-supported but the final Bi-2223 may be more phase pure.

Recently some groups have started to analyze the phase formation of Bi-2223 in situ by synchrotron radiation [50, 111, 226, 227] or neutron diffraction [66]. Bi-2223/Ag monocore tapes undergo the usual heat treatment in a special furnace designed for use in synchrotron or neutron diffraction studies. The most complete results have been obtained by Giannini and coworkers [66], as in their neutron diffraction experiments they were able to make a complete quantitative analysis of seven phases taking part in the Bi-2223 phase formation. These phases are Bi-2201, Bi-2212, Bi-2223, 3221, the 14:24 phase, calcium cuprate and copper oxide. The results support the picture of liquid-assisted phase evolution in the early stages of the processing. Furthermore, these authors were able to analyze the crystalline matter present in the tape and thus they could trace the texture development during sintering. This study shows, in particular, strong evidence for decomposition of the Bi-2212 and subsequent nucleation and growth of Bi-2223. Up to now this is the most informative experiment reported in the literature. However, the studies were performed under ambient atmosphere and with a precursor composition from which the liquid-supported reaction would be expected. More experiments on different reaction routes would be desirable.

5.2.2 Microstructure and Grain Boundaries

A fully reacted Bi-2223 tape contains a lot of different microstructural features which are a conseqence of the reaction kinetics. Larger than micron-sized features such as voids, second phases or misaligned grains can easily found in the optical or scanning electron microscope. However, owing to the short coherence length of the HTSC, submicron-sized features are also of great importance. These defects can only be found in detailed transmission electron microscopy (TEM) studies. TEM analysis is a very sophisticated experimental method. Consequently only few groups are able to perform these experiments. A lot of detailed work has been reported by Babcock and coworkers [11], Eibl [46], Grindatto and coworkers [78] and Holesinger et al. [99]. In these experiments it appeared that second phases are also present on a submicron scale between grains. In addition there is signifcant (10%) Bi-2212 intergrowth in Bi-2223 grains, as Eibl could show [46]. Holesinger [99] could furthermore show that the stoichiometry on a submicron scale is inhomogeneous, identifying submicron 3221 inclusions in particular as deleterious for the supercurrent as they cause a lead depletion in the surounding Bi-2223 phase.

Special attention has to be paid to the properties of the grain boundaries, as already discussed above. Various kinds of grain boundaries can be found in the microstructure of Bi-2223 tapes. Very-low-angle a and b axes grain boundaries are the best candidates for good current transport; however, high-angle grain boundaries and c axis grain contacts are very dominant in the microstructure. Grindatto et al. [78] could show that the most dominant small-angle grain boundaries are small-angle c axis tilt and edge-on c axis tilt boundaries as described in Chap. 2 of this book (see Fig. 2.2). This observation led to the railway switch model of Hensel et al. [95].

The detailed structure of grain boundaries (GBs) in HTSCs has been studied to some extent by Babcock and coworkers [11]. But up to now most of the research has been performed on YBCO bicrystals and thin films. Babcock et al. could show that the GBs in YBCO bicrystals show faceting and dislocations. These features and stoichiometric variations within the GB lead to a new vortex state which influences the current transport over the GB [27]. In particular, in YBCO bicrystals a crossover is observed from strong- to weak-link behavior at angles lower than 10° [90, 92, 93]. Gurevich and Pashitskii [81] proposed a grain boundary model which explains well these experimental results.

Not only are these studies on isolated grain boundaries of basic character, but the particular microstructure of technical HTSCs shows that they are crucial for the understanding and improvement of current transport in these grain-boundary-dominated materials. Initial TEM results on YSZ/CeO_2 buffer layers for YBCO-coated conductors show that they consist of a network of low-angle grain boundaries [277] similar to those in BSCCO conductors.

6. Characterization of Conductors and Bulk HTSCs

In the previous chapter several of the most important characterization methods have already been adressed, but besides microstructural and compositional investigations, physical properties such as current-carrying capability and magnetic behavior have to be examined, too. For certain, the most basic properties of a superconductor are its current-carrying capability and its behavior in an external magnetic field.

6.1 Electromagnetic Characteristics

Usually the first physical quantity which is determined when a superconductor comes out of the preparation process is its critical current. The standard four-probe method is used in every laboratory and I–V curves are measured in order to determine I_c from these curves using the $1\,\mu V$ criterion. But there is more information in the I–V characteristics. Their shape can be evaluated to estimate the influence of defects in a BSCCO conductor on current transport [222]. The slope of the transition (n value) is a measure of the flux dynamics [45]. The most interesting analysis, however, is the investigation of multiple I–V curves for different temperatures and external magnetic fields. These data are used by several groups to determine the ireversibility field H^* [8] and the vortex glass transition [126] of BSCCO conductors or YBCO and Bi-2212 thin films [134, 263]. The results of the scaling analysis, following the original paper of Koch et al. [134], are interpreted as a transition of the flux line lattice from a vortex glass to vortex liquid behavior. The microstructural investigations of grain boundaries as described in the previous chapter are supported by measurements of the transport currents over the GBs. Also, single filaments of high-j_c BSCCO tapes have been analyzed this way, yielding the result that signatures of c axis transport are still present [27].

Transport measurements have the drawback that current and voltage contacts have to be applied. Also, with a macroscopic transport current, mainly intergrain properties are tested. Thus contactless methods such as magnetization and susceptibility experiments are performed, especially when information about the intragrain properties like the flux-pinning characteristics is desired. Magnetization and magnetic-susceptibility measurements using different ac and dc techniques are applied as standard tests by most research groups working on superconductivity. There are, however, groups who specialized in this field, such as [148, 149, 150, 271, 281], to mention only a few. This field is quite sophisticated, though. Therefore a more detailed discussion would be beyond the scope of this book.

6.2 Superconducting Magnetic Levitation

When the first experiments on superconducting magnetic levitation were performed, this effect was still fascinating from a basic point of view and several approaches were made to understand the basic physics behind it [20, 158, 198]. From the point of view of applications superconducting levitation is very interesting as the use of ceramic HTSC frictionless bearings is possible. A huge amount of papers has now been published where levitators, trapped-flux motors, flywheels and other possible applications or even prototypes are presented. The experimental characterization methods are comparable in all cases. The levitation force is measured using a test magnet in a simple setup. In addition, trapped magnetic flux is examined using a scanning Hall probe technique and pinning properties are analyzed by magnetization experiments. The most important progress in this field has been made in sample preparation. In particular, Murakami's group at ISTEC in Japan has succeeded in preparing a mixed-phase RE-123 superconducting ceramic with superior superconducting and mechanical properties [203]. In a recent review Hull summarizes the results on superconducting magnetic levitation [105, 106].

6.3 Imaging of Magnetic Flux in Type II Superconductors

When the lower critical field H_{c1} is reached, magnetic flux penetrates a type II superconductor. The higher the pinning forces, the less flux is able to enter. A hard superconductor, which means it has good pinning properties, thus has a high critical current density. This implies that, as already discussed in Sect. 4.3, flux penetration is directly correlated with the current-carrying capabilities of a superconductor. Consequently, magnetic-flux mapping offers the possibility of measuring important sample properties nondestructively and with spatial resolution. The first experiments were performed using Hall probes and mechanical scanning systems. Later on, magneto-optical imaging (MOI) techniques were established. Hall probe measurements exhibit low spatial resolution but are less difficult to perform. Therefore flux mapping using Hall sensors is used for fast nondestructive testing of large areas or long lengths of superconductors, while MOI is applied for more detailed analysis.

6.3.1 Scanning Hall Probe Experiments

Several years ago Hall probes were widely used to characterize YBCO melt-processed ceramics [62, 161]. Nowadays they are still used for the measurement of trapped-flux magnets [59]. In recent years it became apparent that this method is also very suitable for the measurement of BSCCO conductors. In particular, when a fast nondestructive evaluation of a long length of conductor is necessary, scanning Hall probe measurements are a suitable technique. In this case the self-field of a transport current is measured either along the tape length [17, 241], in order to identify defect areas with reduced critical current density, or across the tape width, which, when an ac current is flowing, provides information as to whether the filaments of a multifilametary tape are coupled [220]. In both cases scanning is necessary only in one dimension and therefore the technique is fairly fast. Its resolution, however, is of the order of a millimeter and thus small defects on a micrometer scale are not detectable.

6.3.2 Magneto-Optical Imaging

For magnetic-flux mapping on a scale of several micrometers magneto-optical imaging has proved to be a very powerful method. Originally used by a group at Chernegolovka to characterize magnetic materials [109, 223], it was soon also applied for the imaging of magnetic flux in high-T_c superconductors. For the latter materials this technique turned out to be very useful. The role of grain boundaries and microcracks as current-limiting defects could be demonstrated successfully [216]. Several groups around the world now use this technique in order to analyze the homogeneity of the flux distribution in BSCCO tapes [244, 132, 68], YBCO-coated conductors [278], melt-textured ceramics [133] and thin films [142]. Correlating these results with the current-carrying capacity and the microstructure as found in an optical or electron microscope enables the researcher to tell which features block or reduce supercurrent transport in the samples examined. Consequently MOI has turned out to be a necessary basic characterization tool in HTSC optimization.

Part III

Phase Formation and Microstructure

7. Preparation of BSCCO Conductors

The powder-in-tube (PIT) process as described in Sect. 3.4.2 was used by the present author to prepare wires and tapes from BSCCO materials. For both Bi-2212 and Bi-2223, green-wire fabrication was identical. Only the thermomechanical processing was significantly different. Therefore the production of green conductors will be treated in one section for both materials.

7.1 Fabrication of Green Wires and Tapes

Following the scheme in Sect. 3.4.2, the processing starts with the treatment of the precursor powders. After some initial experiments with powder preparation [51], commercially available precursors from the companies Merck and Solvay were used. Before filling a calcination step is applied. The silver tubes have to be cleaned carefully before filling. This was done mechanically and by chemical etching using a mixture of one part H_2O_2 and three parts NH_4OH. This mixture etches the silver surface and thus removes remaining oxides and sulfides. The tube is then sealed at one end. When filling with the powder, the silver tube is placed in a hollow cylinder as shown in Fig. 7.1 in order to avoid bending of the tubes during the filling process. The precursors are densified mechanically up to a density of $1.6-2.3\,\mathrm{g/cm^3}$. This corresponds to about 28–41% of the theoretical density of $5.6\,\mathrm{g/cm^3}$. As an intial density of more than 40% does not affect the final density within the processed conductor, as could be shown by Zhang et al. [284], experiments using cold isostatically pressed precursor powders [201] were not continued. The filled silver tube is then transferred into a furnace where it is heated at 400°C for at least two days. Afterwards the evacuated tube is sealed. This procedure reduces the remaining moisture in the assembly, and consequently bubble formation during processing, as observed by other groups [16], is avoided.

Mechanical deformation follows as the subsequent step in the PIT process. Deformation techniques play an important role in achieving good homogeneity of the conductors over long lengths. A detailed consideration of the forces acting during mechanical deformation is given by [84]. The deformation process takes place in several steps. As we did not prepare long lengths of conductors extrusion was not necessary. Profile rolling and wire drawing were used instead as starting steps. Monocore and multifilamentary wires and

86   7. Preparation of BSCCO Conductors

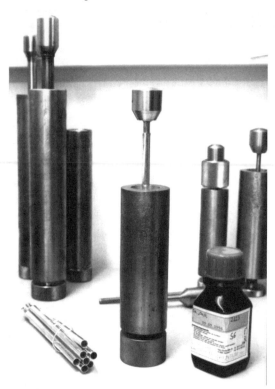

Fig. 7.1. Hollow cylinders used to avoid bending during filling the precursor powders into the silver tubes

tapes were produced in cooperation with Cryoelectra GmbH in Wuppertal. Most of the mechanical processes were performed in the Duisburg laboratory of this company. Tape rolling and pressing between sintering cycles was done at the University of Wuppertal.

Multifilamentary conductors were fabricated by restacking monocore wires into silver tubes and repeated mechanical processing. Drawing and profile rolling and a combination of both techniques were applied. The profile roll consists of square shaped grooves of decreasing size. The forces are quite different in the two processes. The highest force acts in the drawing direction during wire drawing, while in a rolling process the highest force acts normal to the surface of the rolled conductor. Consequently, the typical damage is tearing in the case of wire drawing and cracks in the silver sheath during rolling owing to the asymmetric force distribution. Cross sections of a 4×4 multifilamentary wire prepared using only profile rolling and a 19-filament wire which was fabricated with combined drawing and profile rolling are shown in Figs. 7.2 and 7.3 for comparison. These pictures show that with the latter process a more regular arrangement of the filaments within the silver sheath can be obtained. During drawing, the filaments arrange according to the "magic numbers" $a_n = n^2 + 3n + 1$ (7, 19, 37, 61, etc.). Once the filaments are fixed in their position the conductor can be

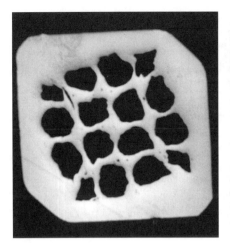
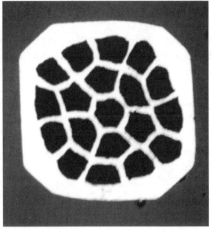

Fig. 7.2. Cross section of a 16-filament wire fabricated by profile rolling

Fig. 7.3. Cross section of a 19-filament wire fabricated by a combination of drawing and profile rolling

profile rolled without changing this arrangement. A further possibility was to use hexagonal or square-shaped monocore wires instead of round ones for restacking into multifilaments. This variation of the geometrical arrangement allows for a higher space filling. In principle, with round wires 82.7% of the space within a round tube can be filled. With a hexagonal arrangement 93% space filling is possible. Only square filaments within a square tube yield 100% filling theoretically. Three possible arrangements were tested experimentally. The highest critical currents with the lowest spread were achieved with the hexagonal arrangement of wire drawing. The results are depicted in Fig. 7.4.

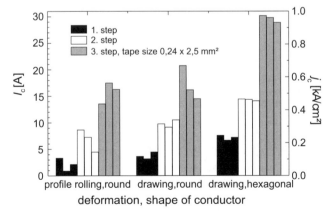

Fig. 7.4. Comparison of different deformation techniques

The final step in mechanical deformation is tape rolling. Experiments showed that there was an optimum tape thickness around 270 μm. For higher and lower thicknesses the critical current density is lower, as depicted in Fig. 7.5. The decrease in j_c with increasing tape thickness is commonly explained by reduced texture of the superconducting core. It is well known from several experiments [216, 163] that the critical current density is higher at the silver–BSCCO interface owing to the enhanced texture of the micrograins. With higher tape thickness more poorly aligned grains within the center of the core are present, which reduce the overall critical current density. The reduction of j_c with decreasing thickness below 270 μm may be caused by an inhomogeneous longitudinal section introduced by sausaging. A more detailed discussion of the influence of the mechanical deformation process is given in [101].

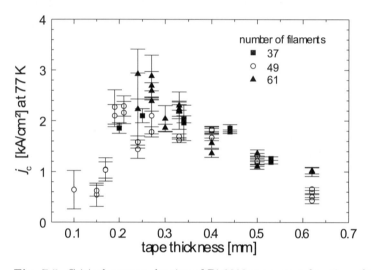

Fig. 7.5. Critical current density of Bi-2223 tapes as a function of thickness after the first sintering step

7.2 Thermal Processing of Bi-2212/Ag Conductors

The procedures for thermal processing of Bi-2212 and Bi-2223 differ strongly. While Bi-2212 can be formed by a melt process, Bi-2223 has to be formed by a solid-state reaction as the compound decomposes irreversibly above its melting point.

7.2.1 Processing Scheme

The initial optimization studies were performed on Bi-2212 monocore wires. A typical temperature–time diagram is shown in Fig. 7.6. There is a whole set of not necessarily independent parameters which have to be optimized. Firstly the maximum temperature T_{top} has to be determined, which has to be higher than the melting temperature of Bi-2212. This melting temperature itself is a function of the atmosphere and the stoichiometry of the precursor powder. Stoichiometric Bi-2212 precursor powders were used throughout the experiments. After a certain dwell time at T_{top} the melt is cooled to T_{sinter}, which lies below the melting temperature. At this point recrystallization sets in. Cooling rates and the dwell time at T_{sinter} also affect the results.

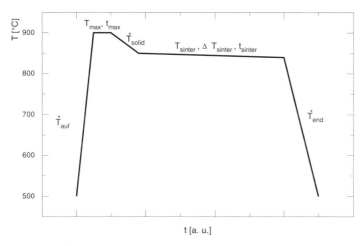

Fig. 7.6. Schematic temperature–time diagram for the processing of Bi-2212 conductors

In order to optimize T_{top} under ambient atmosphere and pressure, T_{sinter} was kept constant at 840°C. The dwell time was varied up to 120 min and the sintering time between 20 h and 120 h. Figure 7.7 shows the critical current and critical current density of monocore and multifilament conductors at 77 K in self-field as a function of the maximum temperature T_{top}. The melting temperature of Bi-2212 under ambient conditions is 880°C. Consequently a critical current can only be detected after processing above this temperature. The monocore conductors show a distinct maximum at 913°C, while the multifilaments show a plateau above 910°C.

7.2.2 Void Formation

The reason for the different behavior of monocore and multifilamentary conductors becomes clear from quench studies. In Fig. 7.8 optical micrographs of

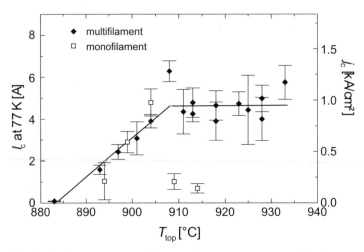

Fig. 7.7. Optimization of the maximum temperature T_{top} for monocore and multifilamentary Bi-2212 conductors

longitudinal sections of Bi-2212 monocore wires are shown. The wires have been heated up to the indicated temperature and subsequently quenched in liquid nitrogen. At 670°C the core still consists of compacted powder. At 880°C the powder melts and starts to compact further, forming voids in the neighborhood of the melt. With increasing maximum temperature the voids coalesce while big voids grow bigger at the expense of smaller ones (Ostwald ripening). At 920°C these cavities are big enough to fill the total cross section of the ceramic core. In the case of a monocore wire this means complete blocking of current transport, while in a multifilament conductor there are still other filaments left which can contribute to the current transport.

There are several reasons for porosity in Bi-2212 conductors [284]:

1. Expanding gases such as N_2 or water vapor.
2. Carbon impurities in the precursor powder, which are released in the form of CO_2.
3. Densification of the core due to melting.

The first possibility has been excluded by the pretreatment at elevated temperature in vacuum.

Concerning the second possibility, as the precursor materials for powder formation contain carbonates a certain amount of carbon remains in the starting powder as impurity. During the preparation process the carbon forms CO_2 and leaves the melt. Depending on the amount of carbon impurities this can even lead to bubble formation in the silver sheath. Zhang and coworkers [284] found that a significant reduction of carbon impurities to below 200 ppm and processing in 100% oxygen stops bubble formation.

7.2 Thermal Processing of Bi-2212/Ag Conductors 91

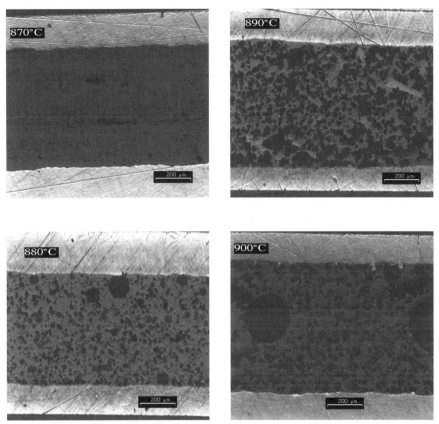

Fig. 7.8. Optical micrographs of Bi-2212/Ag wires quenched from different stages of the sintering process

Concerning the third possibility, the remaining porosity stems from the fact that the packing density in the green wire after mechanical deformation is between 75% and 85%. As the formation of the Bi-2212 superconductor within the silver sheath is a melt process a densification up to theoretical density takes place, leaving up to 25% of porosity.

Porosity is much higher in a wire than in a tape. This might partly be due to a higher densification of the green tape caused be the additional rolling. In addition the silver sheath is much thinner and, owing to its geometrical shape, is less stable in a tape than in a wire. Consequently, collapsing during thermal processing is more likely for the tape geometry. In order to reduce porosity, overpressure processing has been performed in collaboration with the Applied Superconductivity Center of the University of Wisconsin at Madison. The results of these experiments will be presented in detail in Chap. 8.

7.2.3 Microstructure Development

After incongruent melting of the ceramic core the processing temperature is lowered to T_{sinter}, which usually is 840°C. At this point Bi-2212 starts to recrystallize from the melt. Bi-2212 crystals form from alkaline-earth cuprates and copper-free compounds according to

$$\text{Liq.} + (\text{Sr}, \text{Ca})_{14}\text{Cu}_{24}\text{O}_x + \text{Bi}_2(\text{Sr}, \text{Ca})_2\text{O}_x \rightarrow \text{Bi} - 2212, \tag{7.1}$$

$$\text{Liq.} + (\text{Sr}, \text{Ca})\text{CuO}_2 + \text{Bi}_2(\text{Sr}, \text{Ca})_4\text{O}_x \rightarrow \text{Bi} - 2212. \tag{7.2}$$

The velocity of this reaction is quite slow. The maximum critical current density was only reached after 150 h of sintering time. The critical current density as a function of sintering time is shown in Fig. 7.9.

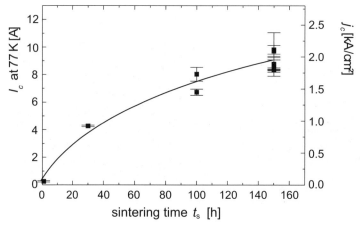

Fig. 7.9. Critical current density as a function of sintering time

The development of the microstructure during the preparation process was studied by quench experiments. Figure 7.10 denotes the stages in the process at which samples were quenched into liquid nitrogen. Electron microscopy was then performed on these samples.

Figure 7.11 shows the ceramic core 6 min after the maximum temperature of 920°C has been reached. Obviously, still not all of the material is molten. Between the melt regions there are still powder particles present. After 30 min at 920°C the material is completely molten as can be seen from Fig. 7.12. Figure 7.13 shows the ceramic core after 100 h of sintering time. The platelet-like microcrystals of Bi-2212 are clearly visible. There is no preferential texture as the sample was a Bi-2212 wire. The crystal growth seems to be complete after 50 h, as no obvious difference in microstructure between wires sintered for 50 h and 100 h could be found.

Nevertheless the critical current density still improves at up to 150 h of sintering time. There might be processes on a much smaller scale along the

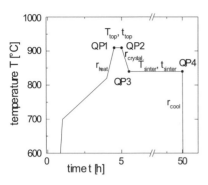

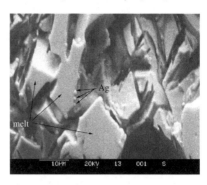

Fig. 7.10. Points in the time-temperature diagram from which samples were quenched

Fig. 7.11. Microstructure of a Bi-2212 monocore wire obtained by electron microscopy shortly after reaching $T_{\mathrm{top}} = 920°\mathrm{C}$ (QP1)

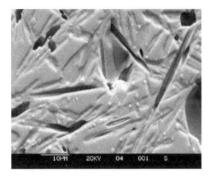

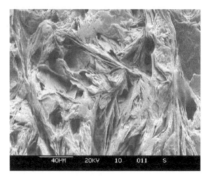

Fig. 7.12. Microstructure of a Bi-2212 monocore wire obtained by electron microscopy after 30 min at maximum temperature (QP2)

Fig. 7.13. Microstructure of a Bi-2212 monocore wire obtained by electron microscopy after completion of the sintering process (QP4)

grain boundaries which are not complete after 50 h. The quench experiments presented here, however, are not exact enough to obtain more detailed results, especially as owing to the Leidenfrost phenomenon the cooling rate in liquid nitrogen is as low as about 600 K/s.

7.3 Thermo-Mechanical Processing of Bi-2223/Ag Tapes

As already discussed to some extent in Sect. 3.4.3, the preparation route of Bi-2223 is more complex than that of Bi-2212. The compound has to be kept well below the melting point, otherwise it decomposes irreversibly. In the beginning of the reaction a small amount of liquid phase is present to support the growth of Bi-2223 grains. The main reaction is a solid-state reaction and mechanical treatment is necessary to enhance grain alignment.

7.3.1 Precursor Powder

As the precursor a commercially available powder (Merck) was used. The production process is spray pyrolysis, which introduces variations in the cation stoichiometry. The company gives the tolerance interval as 3%. As the cation stoichiometry affects the reaction kinetics – the reaction is faster with increasing Ca and Cu content [44, 239, 275] – and the melting temperature, an optimization process has to be performed for each new powder batch. All experiments presented were conducted with one powder batch.

The main compound in the starting powder is Bi-2212, as can be seen from x-ray diffraction in Fig. 7.14. The overall composition of the powder ME7 is summarized in Table 7.1 including the nominal (wanted) and the actual (measured by X-ray analysis) stoichiometry, as given by the supplier. All values are normalized to the copper content.

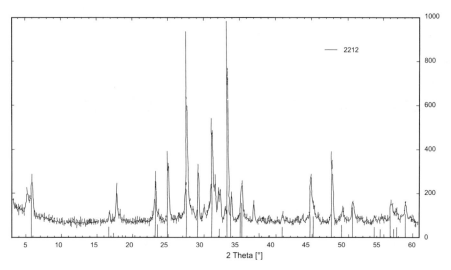

Fig. 7.14. X-ray diffraction analysis of the starting powder ME7. The main compound is Bi-2212

As already mentioned, traces of carbon are the most unwanted impurity. Carbon is known to precipitate at grain boundaries and is also responsible for porosity due to gas development during processing. Thus it has a significant influence on grain connectivity [56, 276]. A carbon content below 0.38 atom % is considered to be low enough [275].

Table 7.1. Elemental analysis of the Bi-2223 precursor ME7 as delivered by Merck. The values are normalized to the copper content

Element	Weight %	Atom %	Actual	Nominal
Lead	6.7	3.8	0.34	0.33
Bismuth	37.4	21.0	1.91	1.80
Strontium	14.7	19.7	1.79	1.87
Calcium	7.4	21.6	1.97	2.00
Copper	17.9	33.0	3.00	3.00
Nitrogen	< 0.05	< 0.42		
Carbon	< 0.05	< 0.49		
Barium	< 0.01	< 0.01		
Iron	< 0.01	< 0.02		
Silicon	< 0.01	< 0.04		
Zirconium	< 0.01	< 0.01		

7.3.2 Number of Processing Steps

A schematic outline of the thermo-mechanical treatment is given in Fig. 7.15. Intermediate rolling or pressing plays a key role in this process route. This mechanical step is necessary to densify the ceramic core and obtain good alignment of the Bi-2223 microcrystals. However, during deformation cracks are introduced, which have to be healed in the subsequent thermal processing. With increasing reaction time the amount of liquid phase available for healing the cracks decreases. Thus after a certain number of thermo-mechanical cycles the critical current density can no longer be improved. This is obvious from Fig. 7.16, where the critical current and current density are plotted as a function of sintering time with an increasing number of mechanical processing steps.

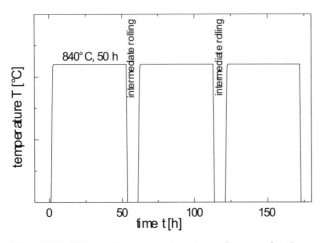

Fig. 7.15. Schematic temperature–time diagram for the preparation of Bi-2223/Ag tapes

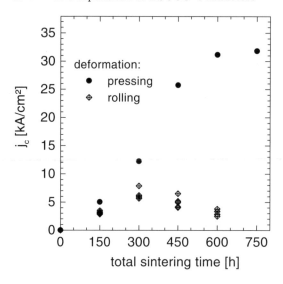

Fig. 7.16. Repeated intermediate mechanical treatment (pressing and rolling) of a Bi-2223/Ag 37-filament tape. The sintering temperature was kept constant at 836°C

In this plot the principal difference between rolling and pressing shows up in the optimum number of intermediate mechanical treatments. While after two rolling steps the critical current density decreases, indicating that the cracks introduced are no longer healed, up to four pressing steps are necessary to reach a plateau. The overall critical current density which can be reached by pressing is much higher than in rolled tapes. The maximum value obtained was $32\,\mathrm{kA/cm^2}$ at 77 K in self-field for pressed tapes.

7.3.3 First Thermal Processing Step

The preparation procedure was optimized successively, starting with the first sintering step. The important parameters are the processing temperature T_1 and the processing time t_1, which are varied in a narrow grid. This is usually a very time-consuming procedure. A gradient furnace was used to obtain the results in a shorter time. The calibration of this furnace was performed very carefully; thus each position inside the chamber has a certain known temperature. Putting samples at different positions within this gradient allows for preparation of several points of the parameter matrix in one step. A temperature interval of 30°C could be covered with one preparation step.

Figure 7.17 shows the critical current as a function of processing temperature in the first step for 19- and 49-filament Bi-2223 tapes. The holding time was 50 h in each experiment. There is a pronounced maximum at 852°C, which corresponds to the furnace temperature Θ. The real temperature T in this furnace is 11°C below the indicated value. Below this temperature I_c decreases slowly; above it the decrease is fast and can be explained by the decomposition of the core material as already discussed. A different number of filaments leads to a different absolute value of I_c but the shape of the curve

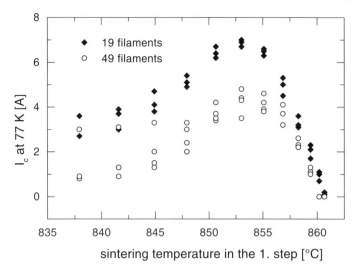

Fig. 7.17. Optimum temperature (furnace temperature) for the first thermal treatment determined in a gradient furnace for Bi-2223/Ag 19- and 49-filament tapes

is similar. For further processing a maximum temperature of $\Theta = 850°C$, corresponding to a real temperature of $T = 839°C$, was chosen. This is slightly below the optimum temperature as determined from the optimization procedure. But taking into account possible deviations of temperature in any given furnace it is safer to choose the lower temperature in order to avoid decomposition of the material at the high-temperature side.

The second parameter to optimize is the duration of the first sintering step. The processing temperature was kept constant at $T = 839°C$. After a certain sintering time I_c and T_c were determined and the samples were analyzed by x-ray diffraction. Figure 7.18 shows the XRD spectra as a function of sintering time. The identification of the x-ray peaks was performed according to [206]. After 10 h of sintering time Bi-2212 is still the dominating phase. Within the first 30 h Bi-2212 reacts with Ca_2PbO_4 and CuO to form Bi-2223. This reaction is quite fast as the plumbates, as a liquid phase, enhance the reaction kinetics. After 30 h most of the liquid is consumed and the reaction becomes very slow, as is typical for a solid-state reaction as described in Sect. 3.4.3. The amount of Bi-2223 phase is already around 50% after 15 h. After 110 h of sintering the Bi-2223 phase content is as high as 94%. But there is still Bi-2212 present in the XRD spectrum.

The Bi-2223 phase content and the critical current I_c are correlated with the sintering time in Fig. 7.19. This plot clearly shows the reaction kinetics. After about 15 h the first nonzero I_c is measured. The phase content of Bi-2223 is 50% and only now is a connected current path possible. The critical current increases fast, together with the Bi-2223 phase content, up to a processing time of about 30 h. The further increase in critical current, as well as

98   7. Preparation of BSCCO Conductors

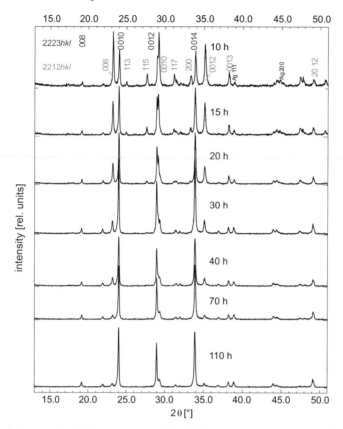

Fig. 7.18. XRD analysis of Bi-2223/Ag tapes processed with increasing sintering time in the first processing step

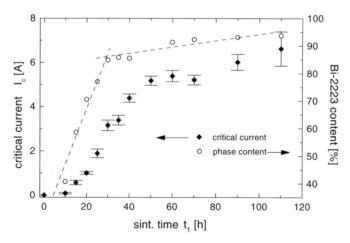

Fig. 7.19. Bi-2223 phase content and critical current of Bi-2223/Ag tapes as a function of sintering time in the first processing step

in phase content, is slow but still has a positive slope at 110 h. This behavior corresponds well to the reaction scheme described in Sect. 3.4.3.

7.3.4 Further Processing Steps

A second thermal processing usually follows after intermediate pressing or rolling. For the determination of the optimum temperature in the second step this intermediate mechanical treatment was omitted. The first processing step was performed in a homogeneous furnace at $\Theta = 847°C$ for 50 h. The second temperature was then again optimized in the gradient furnace. The results are shown in Fig. 7.20. Here the critical current is plotted as a function of the temperature in the second processing step. The results from the optimization of the first step are added for comparison. As can be seen, the optimum temperature in the second thermal processing lies 17°C below the optimum temperature in the first step. However, if an intermediate mechanical process is introduced between the first and the second sintering step the critical current densities in the second step degrade significantly. This indicates that it is of advantage to use a lower sintering temperature in the third stage and apply an intermediate mechanical processing with subsequent sintering at the same temperature in order to heal the cracks introduced by the rolling.

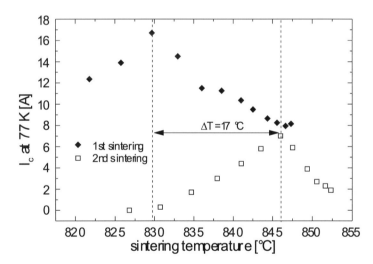

Fig. 7.20. Optimum temperature in the second processing step in comparison with the first sintering process

The optimum temperature in the third step was determined in a comparable manner to the temperatures before. Samples were sintered at $\Theta = 847°C$ two times for 50 h with intermediate rolling. The temperature of the final step was then determined in the gradient furnace as described before. The result is

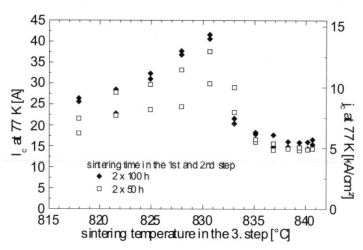

Fig. 7.21. Optimum processing temperature Θ_3 in the third stage of the sintering process

shown in Fig. 7.21. Again the reduced temperature is $\Theta = 830°C$, corresponding to $T = 819°C$ absolute as determined in the second step. The reduction of the processing temperature in the final step leads to significantly enhanced critical current densities, implying some important experimental results, and will therefore be treated separately in Chap. 9.

8. Overpressure Processing of Bi-2212 Conductors

The principal production route of Bi-2212/Ag conductors using the PIT process under ambient conditions has been described in Chap. 7. From these results it became clear that porosity and void formation are a major problem in the fabrication of Bi-2212 wires and partly also in tapes. Reduction of porosity could be achieved by processing in 100% oxygen as reported by Zhang et al. [284]. Void and bubble formation could be reduced by a gas pressure melting process as proposed by [119] and also applied by Reeves et al. to Bi-2212 tapes [231]. Some detailed experiments have therefore been performed on overpressure processing of multifilamentary Bi-2212 wires and tapes in collaboration with the Applied Superconductivity Center of the University of Wisconsin at Madison.

8.1 Void Reduction in Bi-2212/Ag Wires

8.1.1 Processing Scheme

As already mentioned, void formation is stronger in wires than in tape conductors. This may be partly due to the missing tape-rolling step, which introduces further densification. On the other hand the silver sheath of wires is usually thicker than that of tapes; thus collapsing of the sheath during processing is less probable. Furthermore, round or square wires have a geometrically more stable shape than a tape conductor, which also hinders collapsing of the sheath into the remaining voids. As Reeves et al. [231] have already shown, the critical current density of Bi-2212 monocore tapes can be enhanced by about 20% to 30% by using overpressure processing. The same overpressure processing was therefore applied to multifilamentary Bi-2212 wires. The samples were 19-filament wires of Bi-2212 with a nearly square cross section as shown in Fig. 7.3. The wires were pretreated for one day at 700°C in vacuum and for another two days at 835°C in 100% oxygen. The temperatures given here are all absolute temperatures. After this preannealing step the processing was continued using the temperature–time diagram depicted in Fig. 8.1.

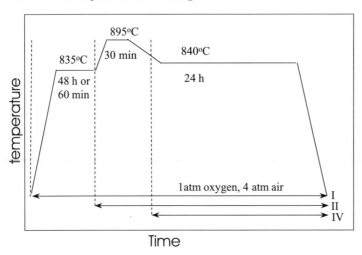

Fig. 8.1. Temperature–time diagram for overpressure processing of Bi-2212 conductors

8.1.2 Critical Current Density

Overpressure was applied by compressing a mixture of oxygen and argon up to 5 atm, yielding 1 atm of oxygen as in the normal-pressure processing plus 4 atm argon. The point when pressure was applied was varied from the very beginning of the process to slightly after melting. The results were compared with normal-pressure-processed samples from the same furnace. The critical current of the samples were measured after processing, in liquid helium at 4.2 K and in self-field. The results are shown in Fig. 8.2 as a histogram. The critical current is plotted as a function of the onset point of overpressure. The Roman numerals refer to the notation in Fig. 8.1. The highest I_c values and the lowest scatter are obtained when overpressure is applied either from the beginning or after the 835°C step. For the latter, the critical current could be doubled from 140 A to 280 A. These values correspond to critical current densities of 48 kA/cm^2 and 96 kA/cm^2, respectively. The cross-sectional area as determined from the green wire in Fig. 7.3 was assumed for calculating j_c, which is a conservative assumption as the superconductor cross section after processing is usually smaller.

8.1.3 Microstructure

Electron microscopy easily reveals the reason for this significant enhancement of the critical current. Figure 8.3 shows an SEM micrograph of the longitudinal section of a 19-filament Bi-2212 wire processed under normal pressure in 100% oxygen. In Fig. 8.4 an SEM image of a longitudinal section of a sample from the same batch but processed with overpressure is depicted.

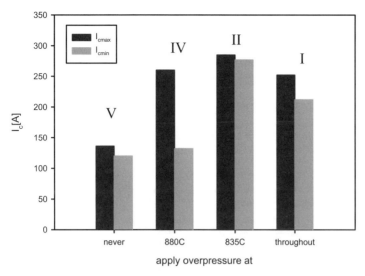

Fig. 8.2. Critical current at 4.2 K in self-field as a function of the onset point of overpressure as denoted in Fig. 8.1, for 19-filament Bi-2212/Ag wires

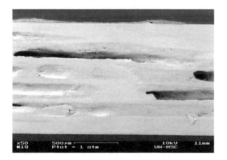 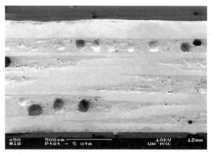

Fig. 8.3. SEM micrograph of a 19-filament Bi-2212 wire processed at 1 atm in pure oxygen

Fig. 8.4. SEM micrograph of a 19-filament Bi-2212 wire processed in 5 atm oxygen

Large voids destroying whole filaments can be seen in Fig. 8.3. The overpressure-processed wire of Fig. 8.4 still has much porosity but the voids are much smaller and more filaments seem to be intact. This implies that overpressure can significantly reduce macroscopic void formation in wires of Bi-2212.

8.2 Overpressure Processing of Bi-2212/Ag Tapes

The success of the overpressure processing of wire conductors of Bi-2212 suggests that conductor geometry plays an important role. Thus it seemed reasonable to apply the same treatment to tape conductors of various thicknesses.

8.2.1 Thickness Dependence of the Critical Current Density

The same preannealing and processing scheme (Fig. 8.1) was used for these experiments. The samples were 15-filament Bi-2212/Ag tapes of various thicknesses. An optical micrograph of a 360 μm thick tape is shown as an example in Fig. 8.5.

Fig. 8.5. Optical micrograph of the cross section of a 15-filament Bi-2212 tape

Tapes with 15 filaments of thickness 160, 260, 330 and 760 μm and 37-filament wires 1.4×1.2 mm square were processed with up to three preanneal steps and under normal pressure or overpressure, throughout or at 835°C, respectively. The critical current of the processed conductors was measured at 4.2 K in self-field. The engineering critical current densities j_e are depicted in Fig. 8.6 as a function of conductor thickness. All samples exhibit a similar thickness dependence of the critical current density; only the position of the maximum varies slightly. For very thick and very thin tapes the critical

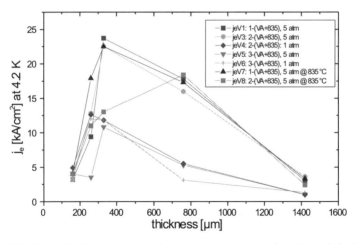

Fig. 8.6. Engineering critical current density as a function of thickness for various preparation processes

current density is lowest. The absolute value of j_e decreases with increasing number of pretreatments.

8.2.2 Magneto-Optical Imaging

In order to find out the reason for the thickness dependence, especially for thinner tapes, magneto-optical imaging was performed on the set of samples denoted V3 in Fig. 8.6. For this purpose the silver sheath of the samples was etched off in a mixture of two parts H_2O_2 and five parts NH_4OH. The samples were then mounted on a copper plate and magneto-optical images were obtained in the apparatus described by [109]. The images are shown in Fig. 8.7.

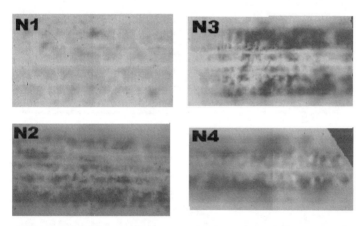

Fig. 8.7. Magneto-optical images of a set of Bi-2212 tapes of thickness 160 μm (N1), 260 μm (N2), 330 μm (N3) and 760 μm (N4). All images were recorded at 15 K and in a magnetic field of 40 mT. The total width of the image area is 2 mm

The thinnest tape is completely penetrated by magnetic flux at a field strength of 40 mT. While samples 2 and 3 show similar flux penetration, sample 3 has large areas where no flux has entered. The amount of flux entry corresponds well to the critical current density, exhibiting the lowest penetration for the highest j_c.

8.2.3 Microstructural Analysis

It was assumed that the origin of the flux penetration behavior lies in the microstructure. Thus the same samples were subsequently examined in an electron microscope (Philips XL30) with EDAX energy dispersive x-ray diffraction (EDX). The results are shown in Fig. 8.8. These images were taken at a magnification of 200 times and are pure or mixed backscattered-electron

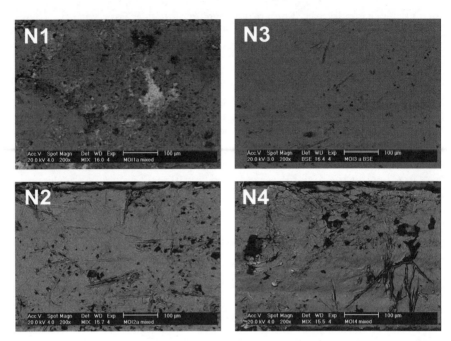

Fig. 8.8. Electron micrographs of a set of Bi-2212 tapes of thickness 160 μm (N1), 260 μm (N2), 330 μm (N3) and 760 μm (N4). All images are pure backscattered-electron images or mixed secondary- and backscattered-electron micrographs

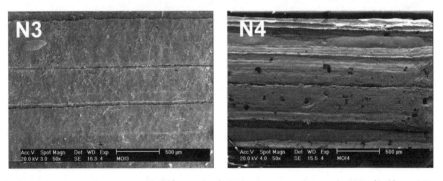

Fig. 8.9. Lower-magnification (50 times) SEM images of samples N3 (*left*) and N4 (*right*) for comparison of overall porosity

images. Sample 1 obviously contains a lot of second phases and has a very fine-grained microstructure. The EDX results show copper-free phase as the main impurity. Sample 2 still contains impurities (copper-free phase) but its microstructure is much more compact. Nearly no second phases are found in sample 3. The microstructure is very compact, as would be expected for a melted and recrystallized material. Sample 4 again has second phases and a lot of porosity.

The secondary-electron images in Fig. 8.9 exhibit the higher overall porosity of sample 4 in comparison with sample 3 more clearly, as they were taken at a lower magnification (50 times).

8.2.4 X-ray Results

In order to obtain more detailed information about the second-phase content of the Bi-2212 tapes x-ray analyses were performed. The overall XRD spectra of all four samples are shown in Fig. 8.10.

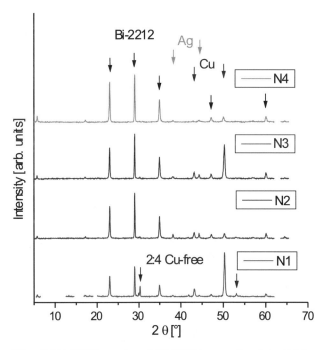

Fig. 8.10. XRD spectra of Bi-2212 tapes of various thickness. The silver reflections stem from the sheath material and the copper peaks from the sample holder

The main Bi-2212 peaks are easily identified. Spurious Ag and Cu reflections stem from the silver sheath and the copper sample holder, respectively. The thinner samples 1 and 2 exhibit weak peaks at around 30°, which can be attributed to the 2:4 copper-free phase. The region between 28° and 32° is shown in a close-up view in Fig. 8.11. The amount of copper-free phase clearly decreases with increasing thickness of the Bi-2212 tape. This is well in correspondence with the EDX results from the electron microscope examinations.

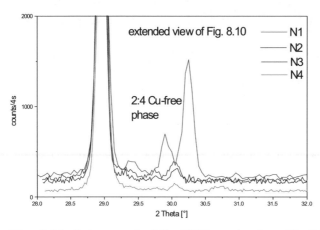

Fig. 8.11. Enlargement of the XRD spectra from Fig. 8.10 around 30° showing the development of the amount of copper-free phase in the Bi-2212 tapes with increasing thickness

8.2.5 Interpretation

The reduction of the critical current density with increasing thickness for the thicker tapes can be understood in terms of incomplete texture and enhanced porosity due to the thickness of the superconducting core. This is a well-known phenomenon in Bi-2223 tapes also. One reason for the poor alignment may be the reduced ratio of silver/Bi-2212 interface as compared with thinner tapes. It is well known from the literature [283] that the degree of texture and j_c are much higher at the interface.

The reason for the j_c reduction in the thinner tapes is the much higher content of impurity phases. This huge amount of second phases, however, is not well understood. One possible explanation might be that the higher interface-to-volume ratio may reduce the optimum sintering temperature. This would imply that the processing temperature used was too high, thus leading to enhanced second-phase production. But this could not be confirmed in differential thermal analysis (DTA) measurements of the tapes. Another possibility is that thinner silver sheaths have a higher oxygen permeability and thus the oxygen concentration within the ceramic core might be changed, consequently changing the phase equilibrium for the optimum reaction. The experimental verification of this hypothesis is not easy. The most probable explanation, however, is the incomplete redistribution of the liquid during melting. Owing to the restricted geometry in the thinner tapes the incongruently melted liquid cannot be redistributed efficiently. This leads to a locally inhomogeneous, nonstoichiometric distribution of the components and thus to an enhanced amount of second phases.

9. Processing of Bi-2223/Ag Tapes at Reduced Final Temperature

As already mentioned in Sect. 7.3, a lower value of the final sintering temperature was found to be useful for the processing of Bi-2223/Ag tapes [85]. This has also been reported by Wang et al. and Zeimitz et al.[265, 280], who apply a two-stage sintering process with intermediate rolling. The experiments presented here were performed with different two- and three-stage processes. The best results were obtained with a three-stage sintering process and a reduced sintering temperature in the final step.

9.1 Processing Schemes

Three different processing schemes were applied. Schematic outlines of these processes are given in Fig. 9.1. All processes have in common one heat treatment at $T = 839°C$ with an intermediate rolling step. Processes I and III consist of a second step at 839°C with subsequent cooling to room temperature. Process II is comparable to the two-stage process applied by Zeimetz

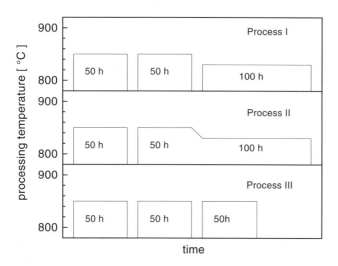

Fig. 9.1. Different sintering processes for Bi-2223/Ag tapes

et al. [280]. The final step in process I takes place at reduced temperature, while in process III the final processing is done at 639°C too. No intermediate rolling was performed between second and last steps.

9.2 Critical Current Density

The lowest critical current density was achieved when all three processing steps were performed at the same temperature. A somewhat higher j_c was achieved when the processing temperature in the third step was reduced without intermediate cooling. This corresponds well with results reported in [280]. The most significant enhancement, however, was obtained on cooling to room temperature between the second and third sintering steps. Typical results for the three processing options are listed in Table 9.1.

Table 9.1. Critical current and relative peak intensity for the BSCCO compounds as obtained from XRD analysis for the different processing schemes

Process	I_c (A)	j_c (A/cm^2)	RPI Bi-2201	RPI Bi-2212	RPI Bi-2223	RPI Bi-3221
I	22.2	15400	0	8.8	90.3	0.9
II	10.0	6900	0	6.5	91.7	1.8
III	7.4	5100	2.0	5.0	90.1	2.9

For several samples processed with variation I the critical current density was measured before and after the third processing step. The results are depicted in Fig. 9.2, as I_c before the last step as a function of I_c after the final step. The data were obtained from a variety of multifilamentary tapes with 19, 37 and 49 filaments. Some results on Ag/Au-clad tapes are also added in the plot. An overall enhancement of a factor of 2.9 is evident from the results.

9.3 X-ray Analysis

In order to obtain information about the reason for the significant j_c improvement, x-ray diffraction (XRD) analysis was performed. The XRD spectra exhibit a change in content of the various BSCCO compounds. Figure 9.3 shows the relative peak intensities (RPIs) for Bi-2201, Bi-3221, Bi-2223 and Bi-2212 for tapes with and without the 819°C processing step. The RPI is not a quantitative but only a qualitative measure of the real content of the respective phase. It is, however, sufficient for comparing the phase development of the different compounds. The RPIs are listed in Table 9.1 for all four

9.3 X-ray Analysis   111

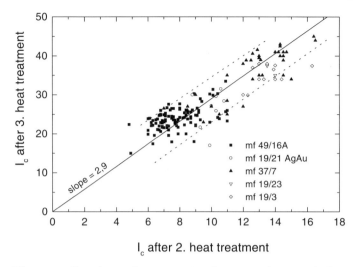

Fig. 9.2. Correlation between critical current densities before and after the last sintering step

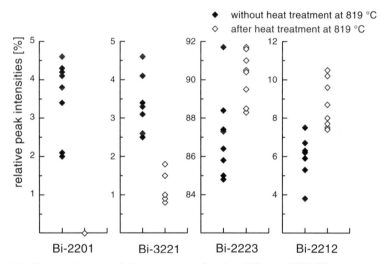

Fig. 9.3. Determination of phase content for the different BSCCO compounds by x-ray diffraction before and after the 819°C step

BSCCO compounds and all three processing schemes. Within the experimental accuracy Bi-2201 is reduced to zero percent when the lower final sintering temperature is introduced. This was also observed by Zeimitz and coworkers [280]. Furthermore we observe an enhancement of the Bi-2212 content and reduction of Bi-3221 in the sequence from process III to process I. The greatest change takes place between process II and process I. Bi-3221 is reduced by a factor of two upon introducing intermediate cooling to room temperature, while the Bi-2212 content rises to 8.8% RPI.

Thus reduction of the final processing temperature reduces the Bi-2201 content, but with intermediate cooling to room temperature in addition, the Bi-3221 content is reduced, too.

9.4 Microstructural Examination

As compositional changes of less than 10% have such a significant influence on the critical current density it is crucial to know where the second phases can be found in the microstructure. Thus scanning electron microscopy was used to study the microstructural features of the above samples. Figures 9.4 and 9.5 show a set of SEM micrographs for samples prepared with all three processing schemes.

The top part of Fig. 9.4 shows the microstructure of a sample before the last processing step. A large amount of Bi-2201 can be found between the grains. The bottom picture exhibits the microstructure of a sample processed with variation II, i.e. without intermediate cooling. No Bi-2201 should be present according to the XRD analysis. But there are still second phases between grains. The length scale, however, is reduced (note the different magnification). The dimensions of these second phases were too small to obtain reliable EDX data. Comparing the micrographs with the XRD data, however, suggests that there is Bi-3221 between the microcrystals.

Figure 9.5 shows two micrographs of samples prepared with intermediate cooling (process I), again with two different magnifications. There is obviously much less second phase present between the grains.

Solidified liquid phases between grains as a by-product of incomplete conversion of the precursor to Bi-2223 have also been reported by Parrell et al. [214]. The results presented here, however, give evidence that it is possible to remove most of these "liquid phases" by the proposed thermo-mechanical processing.

9.5 Ac Susceptibility Results

Results from ac susceptibility measurements [80] give further indications that the grain connectivity is improved by the removal of second phases from the

9.5 Ac Susceptibility Results

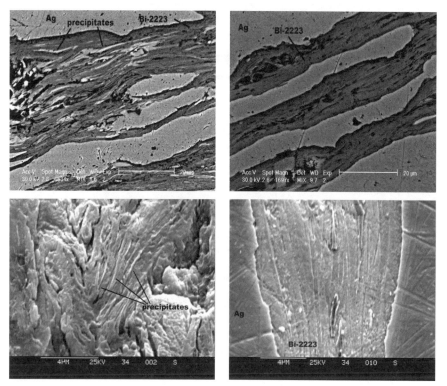

Fig. 9.4. SEM micrograph of a 37-filament Bi-2223/Ag tape before (*top*) and after (*bottom*) final sintering without intermediate cooling. Note the different length scales

Fig. 9.5. SEM micrographs of two 37-filament Bi-2223/Ag tapes after final processing. Note the different length scales on the two pictures

grain boundaries. Figure 9.6 shows results of experiments performed before (top) and after (bottom) the final sintering step.

Before final processing the Bi-2223 tape exhibits a weak screening effect at T_c and additional screening at 98 K. The diamagnetic signal at lower temperatures vanishes with increasing ac field amplitude indicating that the corresponding screening currents are destroyed by the ac magnetic field. After final sintering the tape exhibits good screening behavior and the screening currents are only weakened, not completely destroyed, by an ac field of the same field strength as in the previous experiment.

These results indicate that before final sintering most of the screening takes place within the grains. Only at lower temperatures is a macroscopic screening current present, which, however, can be easily destroyed by higher ac field amplitudes. After final sintering a macroscopic screening current flows at T_c, which is affected but not completely destroyed by the ac field strength.

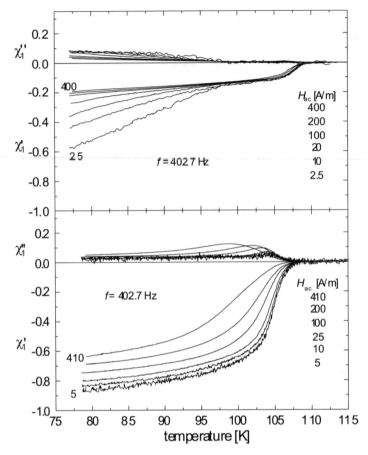

Fig. 9.6. Ac susceptibility (real and imaginary parts) of a Bi-2223/Ag tape before (*top*) and after (*bottom*) the final sintering step of process I as a function of temperature and for various ac field amplitudes

In addition, in the latter case the imaginary part, which is a measure of the ac losses, shows much smaller peaks as compared with the first experiment, implying that the intergrain losses are much lower.

9.6 Discussion

The different results of processes II and III can be well understood, as discussed in Sect. 3.4.3. The results of process II are a confirmation of the results reported by Wang et al. [265] and can be explained by the complete removal of the Bi-2201 phase between grains due to the reduced sintering temperature in the final step. The top of Fig. 9.4 exhibits a similar microstructure to that

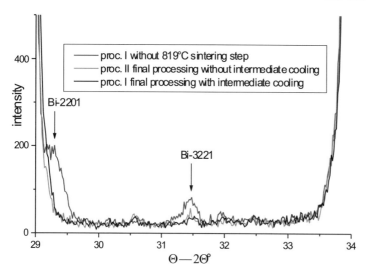

Fig. 9.7. Section of the XRD spectra for samples from process I through II showing the reduction of the Bi- 2201 and Bi-3221 peaks

reported by Wang et al. [265], the light-gray phase being Bi-2201. Concerning the intermediate cooling step (process I), there seems to be a contradiction with the results of Wang et al. [264] and Däumling et al. [41]. Both sets of authors report the enhanced formation of the Bi-3221 phase at temperatures below 800°C.

A tentative hypothesis, which is supported by extra reflections in the XRD spectra, is the following. There is still a small amount of alkaline-earth cuprates in this final stage of the process. A more complete consumption of Bi-3221 and these alkaline-earth cuprates can only take place via a complicated mechanism of decomposition and back reaction of these two phases to Bi-2212 and Bi-2223 . This reaction might be favored by crossing the Bi-3221 phase region, thus consuming more alkaline-earth cuprates. A further hint could be the fact that the reduction of Bi-3221 is accompanied by an enhanced amount of Bi-2212 in the XRD results. For clarification a section of the XRD spectra of samples from processes I through III is given in Fig. 9.7. The reduction of Bi-2201 and Bi-3221 can be clearly seen in this plot.

10. Preliminary Results on YBCO-Coated Conductors

The experiments on YBCO-coated conductors have been started very recently. Consequently only preliminary results can be reported here. The principal preparation process of $YBa_2Cu_3O_{7-\delta}$-coated conductors on biaxially textured metallic substrates has been described in Sect. 3.4.4. In the experiments reported here, different metals were used as substrate materials and all buffer layers as well as the YBCO films were deposited using sputter techniques.

10.1 Metallurgy of the Metallic Substrates

Silver would be the material of first choice as a substrate for YBCO because it neither oxidizes nor reacts with the superconductor. Silver, however, is the only fcc metal known not to form the necessary cube texture easily [70]. It has, rather, to be hot rolled or preheated [253] in order to induce the appropriate texture. Thus the initial experiments on silver were not pursued further. Instead, nickel was used as a base material as it is well known to form a good cube texture easily.

10.1.1 Recrystallization Procedure

The recrystallization of nickel is well known from the literature [200] so the first experiments were performed with this metal. However, nickel is ferromagnetic and thus not suitable for superconducting applications in ac fields owing to the high hysteresis losses. Some nickel alloys, especially Ni–Cu, are known to form good cube texture, too [71]. When choosing an appropriate alloy it is necessary to consider the nickel content at which the alloy is no longer ferromagnetic and the requirement that the lattice constant g does not have too large a mismatch with that of the desired buffer layers and YBCO, in order to enable epitaxy. Figure 10.1 shows the Ni content at which some alloys become ferromagnetic and their lattice constant g in Å. As can be seen from this figure, there are several good candidates. Experimental results on some alloys such as Ni–V or Ni–Cr are known from recent literature [218]. In the experiments presented here pure nickel, Ni–Cu and a Ni–Cu-composite were used.

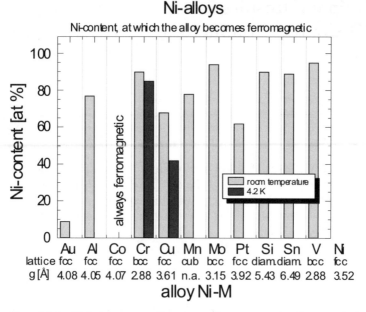

Fig. 10.1. Nickel content of some nonferromagnetic alloys and their lattice constants

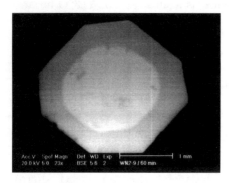

Fig. 10.2. Backscattered-electron image of a profile-rolled Ni–Cu composite wire

Ni–Cu was produced as a composite material. A copper rod was introduced into a nickel tube. Nickel of 99% or 99.98% purity was used. The copper was of OFHC quality. This composite rod was then heated to achieve diffusion bonding of the two metals. The composite was subsequently profile rolled to a cross section of 7 mm and then tape rolled to a thickness between 200 and 270 μm. An electron micrograph of a cross section of a profile-rolled Ni–Cu wire is shown in Fig. 10.2 as a backscattered-electron (BSE) image, which is sensitive to the type of material. The difference in contrast shows the copper core surounded by nickel.

A longitudinal section of a composite tape of 200 μm thickness is shown in Fig. 10.3. An EDX line scan has been performed across the longitudinal

10.1 Metallurgy of the Metallic Substrates   119

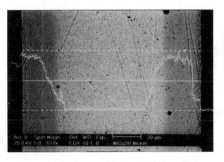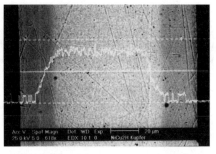

Fig. 10.3. BSE images with EDX line scans for nickel and copper (*left* and *right*, respectively) across a longitudinal section of a composite tape

section for both metals. The results are shown as fine white lines on the electron micrographs (BSE). An interdiffusion zone of a few microns is clearly visible. This material has the advantage that after recrystallization it is mechanically more stable than pure nickel [245]. However, the ratio of copper to nickel used was not high enough to obtain a nonmagnetic composite.

Recrystallization of nickel and Ni–Cu composite tapes was performed in a vacuum of 10^{-4} to 10^{-5} mbar at temperatures between 875°C and 1100°C. Annealing times between five minutes and ten hours were chosen. Figure 10.4 shows the mean grain size of the recrystallized tapes as a function of annealing time for recrystallization temperatures of 900°C and 1000°C.

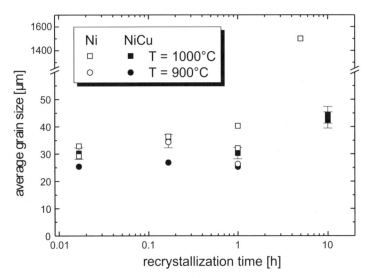

Fig. 10.4. Grain size after recrystallization of Ni and Ni–Cu composite tapes as a function of annealing time at 900°C and 1000°C

The plot shows that at 1000°C the grains grow faster with time in pure nickel than in the composite material. It is an additional advantage of the composite and of alloys in general that grain growth is slower. An overall grain size of around 20 µm would be ideal, because one misoriented grain of size 100 µm or more could completely interrupt the current path in the overlying superconductor.

Grain growth is even faster in 99.98% nickel, as can be seen from polarized-light microscopy. Optical micrographs of nickel and Ni–Cu tapes are depicted in Figs. 10.5 and 10.6. The nickel tape, which has been recrystallized at 900°C, for half an hour has very large grains which are only visible as grains at a low magnification of 50 times. The composite tape, which has been annealed at 900°C for 45 min, exhibits many small grains at a magnification of 200 times. The mean grain size is 30 to 50 µm and the largest grains have a size of 100 µm. This substrate therefore offers a reasonable average grain size for the deposition of buffer layers and YBCO.

Fig. 10.5. Optical micrograph under polarized light of the surface of a recrystallized nickel tape at a magnification of 50×

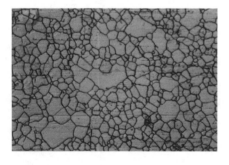

Fig. 10.6. Optical micrograph under polarized light of the surface of a recrystallized Ni–Cu composite tape at a magnification of 200×

10.1.2 Texture Analysis

A qualitative analysis of the texture of the grains can be performed in the optical microscope under polarized light and with differential interference

10.1 Metallurgy of the Metallic Substrates

Fig. 10.7. Surface of a Ni–Cu composite tape under polarized light with DIC after 10 min annealing at 900°C

contrast (DIC). For the latter method Nicol prisms are introduced into the beam path in order to superpose light from different positions of the sample. Height differences of these positions can be resolved down to $\lambda/100$ and better. It is, however, not possible to determine the absolute orientation of a grain. Figure 10.7 shows an optical micrograph of a composite tape which has been annealed for only 10 min at 900°C. There are two main differences from Fig. 10.6. The grains are much smaller and exhibit multiple colors, depicted here on a gray scale. The colors of the grains contain two types of information. Firstly, the grains have many different orientations, thus changing the polarization direction of the light differently. Secondly, the differently oriented grains have been polished to different heights in the previously applied electropolishing process, and thus the DIC shows this height difference as different colors. After only 10 min of annealing time the sample is poorly textured, which appears as different colors in the metallurgical microscope. This means that by using metallurgical microscopy an initial qualitative determination of texture is possible.

In order to obtain quantitative and spatially resolved evidence of the texture of the grains, electron backscatter diffraction (EBSD), also called orientational image mapping [4, 230], has been applied. This method allows one to calculated the exact orientation of each grain and the angles of the grain boundaries by measuring the Kikuchi lines of a backscattered electron beam at many points of a sample. The nickel tapes, however, appeared to have grains too large to apply this method. The results for the Ni–Cu composite tape are shown in Fig. 10.8. [1]

The gray scale in the orientational image map denotes the orientations of the grains. Most of the grains have an 001 orientiation (dark gray), which means that this tape exhibits a good cube texture. In addition, the thick black lines in Fig. 10.8 denote grain boundaries with angles higher than 15° and the thin gray lines mark grain boundary angles between 5° and 15°. Obviously this sample contains only a small amount of high-angle grain boundaries.

---

[1] The EBSD experiments were performed by Dr. E. Bischoff of the MPI für Metallforschung at Stuttgart.

122   10. Preliminary Results on YBCO-Coated Conductors

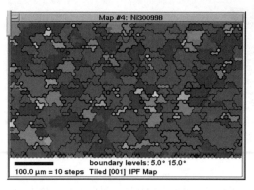
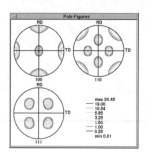

Fig. 10.8. Spatially resolved texture analysis of a Ni–Cu composite tape determined by EBSD (electron backscatter diffraction), and pole figures for the different crystal directions. The 111 pole figure exhibits good cube texture

More detailed information about the texture can be obtained from the pole figure, which is also depicted in Fig. 10.8. The 111 pole figure clearly shows the typical fourfold symmetry of the cube texture, which was the goal of the recrystallization treatment. The inverse pole figures, which are not depicted here, give a impression of the mosaic spread of the orientation. The 001 texture shows a half width of around 12°. This means that the metal substrate is reasonably aligned in plane and out of plane and could serve as a basis for the buffer layers and superconducting coating.

10.2 Buffer Layers

As an initial layer composition CeO_2 + YSZ + CeO_2 was used as it has been reported to have the optimum properties by the ORNL group [75]. As the deposition method a high-frequency magnetron sputter technique was applied. A detailed description of the apparatus and method can be found in [137].

10.2.1 Buffer Layers on Nickel Substrates

Using the parameters given by Qing et al. [228], the first CeO_2 layer was deposited in a forming gas (95% Ar + 5% H_2) atmosphere at 5×10^{-5} mbar. The substrate heater was regulated to a temperature of 360°C. The exact temperature at the substrate is not known but is estimated to be 300°C from previous experiments [251]. According to [228] this should be the optimum deposition temperature for the first buffer layer. The deposition rate was 130 nm/h as measured with a quartz balance. After three hours a 390 nm thick CeO_2 film was obtained. The film was very smooth, as can be seen from electron microscopy (Fig. 10.9). However, the grains of the nickel substrate are still visible. Also, traces of the rolling process can be seen in the morphology of the substrate.

A second layer of YSZ was deposited on top of the CeO_2 layer. Apart from the heater temperature, all deposition parameters were the same. The heater temperature was chosen to be 950°C, which should correspond to a substrate temperature of 800°C. With these parameters the deposition rate was 100 nm/h. After three hours a 300 nm film was produced. An SEM image of this layer is shown in Fig. 10.10. The morphology of the YSZ layer is very interesting. The grain boundaries of the nickel substrate are still visible. Nevertheless the YSZ film looks very different on different grains. The grain in the left corner of the image seems to show traces of the rolling process but is fairly smooth. The other two grains exhibit island growth and incomplete coverage of the surface. This can be explained by a more three-dimensional growth process. A third buffer layer of CeO_2 was deposited on top of the YSZ. The morphology very much resembles that of the YSZ layer and is therefore not explicitly shown.

From these results two conclusions can be drawn. The different morphologies of the YSZ layer on different grains of the nickel substrate might be a consequence of the nonuniform orientation of the nickel grains. Because these grains have different crystal planes parallel to the surface the epitaxy

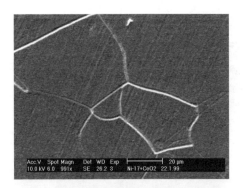

Fig. 10.9. SEM micrograph of a CeO_2 buffer layer on a recrystallized nickel tape

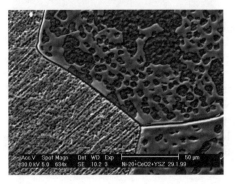

Fig. 10.10. SEM micrograph of a CeO_2–YSZ buffer multilayer on a nickel substrate

conditions change strongly. The three-dimensional growth is a hint that the substrate temperature during deposition might have been too high. Further experiments to clarify these assumption have been conducted on Ni–Cu composite tapes.

10.2.2 Buffer Layers on Ni–Cu Composite Tapes

The CeO_2 and YSZ buffer was deposited on top of the Ni–Cu composite material at 750°C, corresponding to a substrate temperature of 600°C, again in a forming-gas atmosphere. The thicknesses of the films were 390 nm for CeO_2 and 300 nm for YSZ, respectively. Optical microscopy reveals that under these deposition conditions the growth behavior of the films has changed. Figure 10.11 shows the surface of the YSZ top layer at a magnification of 1000 times. No growth islands can be found at this magnification. Only the substrate morphology (grain boundaries, rolling traces) is visible. Further examinations of these films have to be performed especially texture analysis and x-ray diffraction.

Fig. 10.11. Optical micrograph of a 390 nm CeO_2 and 300 nm YSZ buffer multilayer on a Ni–Cu composite tape at a magnification of 1000×

10.3 Y–Ba–Cu–O Coating

For the deposition of the superconductor, high-oxygen-pressure dc sputtering was used. This method is described in detail in [164]. Originally it was applied to produce high-quality films on single-crystal substrates such as MgO, lanthanum aluminate (LAO) or sapphire. With the flexible metal substrates additional difficulties arise. In particular, maintaining good thermal contact during deposition is not very easy. Nevertheless an initial success could be achieved in depositing a 300 nm YBCO film on cube-textured nickel + CeO_2 with a transition temperature of 86 K. The superconducting transition was measured inductively as described elsewhere [65]. The result is depicted in Fig. 10.12.

10.3 Y–Ba–Cu–O Coating

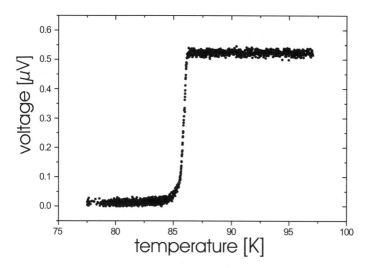

Fig. 10.12. Superconducting transition of an YBCO film on textured nickel and buffer layers measured inductively

A superconducting transition is clearly visible in the inductive signal. Also the width of the transition is fairly small and T_c is about 86 K. The measurement of the transport critical current density of this film yielded reasonable results of j_c of about $0.1\,\mathrm{MA/cm^2}$ at 77 K. Further optimization is necessary in all respects. Reproducibly textured, nonmagnetic alloys have to be produced and the buffer layers have to be optimized. With optimized substrates the growth conditions for the superconductor can be expected to be better. In addition different deposition techniques like co-evaporation have to be considered.

Part IV

Electromagnetic Properties

11. Magnetotransport and Vortex Dynamics

The electromagnetic behavior of HTSCs, in particular the magnetic-field dependence of the critical current density, contains important information about the grain connectivity and flux-pinning properties of a superconductor. In Fig. 2.1 the critical current densities of various superconductors have been depicted as a function of an external magnetic field. From this plot it is clear that the magnetotransport properties are relevant for possible applications, as already discussed in the introduction. Moreover, they give valuable insight into the basic physical properties of flux dynamics in these materials.

11.1 Magnetic-Field Dependence of the Critical Current Density

Experiments have been conducted on monocore tapes and wires of Bi-2212 and Bi-2223. The I–V characteristics have been measured at temperatures between 4.2 K and 100 K in magnetic fields up to 8 T. The tapes were examined in magnetic fields parallel to ($B_\parallel$) and perpendicular to ($B_\perp$) the tape plane. The critical current density was determined from the critical current divided by the superconductor cross section, where the critical current was measured by the four-point technique using a 1 µV criterion. The results for a Bi-2223 tape with $j_c = 32\,000$ A/cm^2 at 77 K in self-field are shown in Fig. 11.1 for both tape–field orientations.

Three different regimes can be distinguished in the field dependence of j_c, their extension varying with temperature.

- *Regime I.* In small magnetic fields appreciably below one tesla, j_c decreases strongly with magnetic field. Weak links decouple in this low-field regime; consequently these grain boundaries no longer contribute to the current transport. Thus the effective area penetrated by the supercurrent is reduced [47].
- *Regime II.* Current flows percolatively across strongly coupled grain boundaries. Vortices are pinned efficiently [102].
- *Regime III.* The critical current density decreases again as vortices start to move through the superconductor under the influence of the Lorentz force and thus dissipate energy. The field at which vortex motion sets in

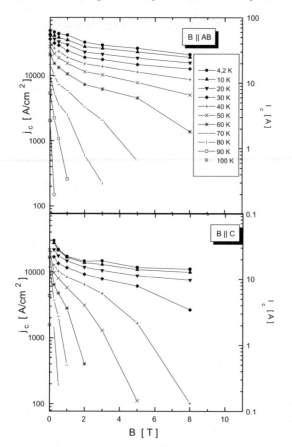

Fig. 11.1. $j_c(B)$ for different temperatures in $B_\parallel$ (*top*) and $B_\perp$ (*bottom*)

is called the *irreversibility field* H_{irr} and strongly depends on temperature and field orientation. The physical origin of this field is controversial in the literature. Two possible explanations are enhanced thermally activated flux creep [127] and a phase transition in the flux line lattice from vortex glass to vortex liquid [168].

As the irreversibility field represents a limit on useful applications of the HTSCs, a physical understanding of its origin is crucial in order to possibly achieve a shift of this limit to higher fields and temperatures.

11.2 Scaling of the *I–V* Characteristics

One possibility to obtain more information about the degradation of the critical current density in magnetic fields is the analysis of the current–voltage characteristics. Figure 11.2 shows *I–V* curves at 3 T for various temperatures, for the same tape as in Fig. 11.1 in a double-log plot. A scaling analysis can

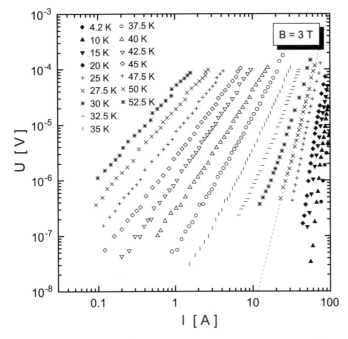

Fig. 11.2. Current–voltage characteristics of a Bi-2223/Ag tape in a 3 T parallel field at various temperatures

be performed with these curves comparable to that in [127, 168, 169]. The basic assumption is that a transition temperature exists at which the pinning correlation length ξ_p diverges like

$$\xi_p \sim |T - T_g|^{-\nu}. \tag{11.1}$$

T_g is determined from the change of the curvature of the I–V curves. From this a scaling of the critical current density and the resistivity follows [134]:

$$j_{\text{scal}} \sim \frac{j\xi_p^{d-1}\phi_0}{k_B T}, \tag{11.2}$$

$$\rho_{\text{scal}} \sim \rho\xi_p^{d-2-z}. \tag{11.3}$$

In these equation ν is the static and z the dynamic exponent. The quantity d denotes the dimensionality of the system. With

$$E = \frac{j}{\rho} \tag{11.4}$$

a scaling of the E–j curves, and correspondingly the I–V curves, follows:

$$E(j) = j\xi_p^{d-2-z} \bar{E}_\pm \left(j\xi_p^{d-1} \frac{\phi_0}{k_B T} \right), \tag{11.5}$$

with

$$\bar{E}_+(j_{\text{scal}}) \to \text{const.} \quad \text{for} \quad T \gg T_{\text{g}}, \tag{11.6}$$

$$\bar{E}_-(j_{\text{scal}}) \to \exp\left(\frac{a}{j_{\text{scal}}}\right) \quad \text{for} \quad T < T_{\text{g}}. \tag{11.7}$$

All E–j characteristics collapse onto two universal master curves E_+ and E_-. Using the dimensionality $d = 3$ the axes in the double-log plot scale as follows:

$$j_{\text{scal}} = \frac{I}{|T - T_{\text{g}}|^{2\nu}}, \tag{11.8}$$

$$\rho_{\text{scal}} = \frac{U}{I\,|T - T_{\text{g}}|^{\nu(z-1)}}. \tag{11.9}$$

The results of the scaling analysis are depicted in Fig. 11.3 for magnetic fields of 2 T and 5 T. The transition temperatures are 58.7 K and 50.4 K, respectively. The exponents obtained, $\nu = 0.9$ and $z = 9.7$, agree well with the values reported in the literature [127, 168]. The analysis works over a broad temperature range, which indicates that there is a smooth transition rather than a sharp phase transition. It is remarkable, that the scaling works best with a dimensionality of three instead of two as would be expected from a strongly two-dimensional superconductor like Bi-2223. This may be a consequence of the multiply connected microstructure of the superconducting core of these tapes. That means that the ceramic core consists of microcrystals, which are – even if well textured – more or less weakly connected. This leads to a percolative current path, which occupies more than two dimensions.

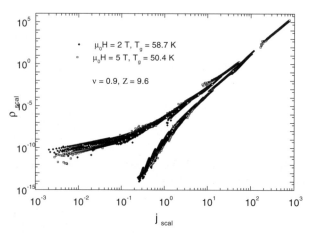

Fig. 11.3. Scaling of the current–voltage characteristics of a Bi-2223/Ag tape in magnetic fields of 2 T and 5 T

11.3 Critical-Current Anisotropy

The scaling behavior can be considered as very universal; thus it is hard to make any statements about a phase transition in the flux line lattice. Therefore a further quantity, the anisotropy of the critical current density with respect to different magnetic field directions, is taken into account. From Fig. 11.1 it is clear that the degradation of j_c in an external magnetic field is much stronger if the field direction is oriented perpendicular to the tape plane, and this degradation and sets in at lower temperatures. This anisotropy results from the special nature of the flux lines in HTSCs as described in Sect. 4.3. In magnetic fields perpendicular to the Cu–O_2 planes, single pancake vortices have to be pinned in each plane. In parallel magnetic fields flux lines are arranged between these planes and are thus strongly pinned intrinsically. Owing to the preparation process of Bi-2223/Ag tapes, as discussed in Chap. 7, the microcrystals within the ceramic core are mainly aligned with their crystal c axes normal to the tape plane. Thus an external magnetic field oriented normal to the tape plane is also oriented parallel to the c axes of the microcrystals. But the texture is not perfect; hence some of these grains are misaligned with respect to this texture. A sketch of this situation is shown in Fig. 11.4.

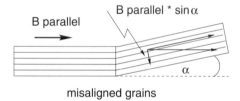

Fig. 11.4. Schematic outline of the definition of the misalignment angle between individual grains in the superconducting core of a Bi-2223 tape

Even if an external magnetic field is applied parallel to the tape plane there is still a component acting parallel to the c axes owing to this misalignment. This component can be taken into account by multiplying $B_\parallel$ with the sine of the misalignment angle α. The reduced presentation of I_c versus $B_\perp$ and $B_\parallel \sin \alpha$ as shown in Fig. 11.5 for $T = 77\,\text{K}$ and $T = 4.2\,\text{K}$ provides values for α. This method is used by several researchers to determine the average misalignment of Bi-2223 tapes [130].

In Fig. 11.6 this angle is shown as a function of temperature. There is evidently a strong increase in α at temperatures above 50 K, which is in contradiction with statements made in [242]. This cannot be attributed to grain misalignment alone, as this is a material constant and should not depend on temperature. It will therefore be called the *magnetic angle* in the following.

As the magnetic angle of the microcrystals is supposed to be temperature-independent there must be an additional mechanism responsible for the temperature dependence of the critical-current anisotropy. We shall discuss two possibilities:

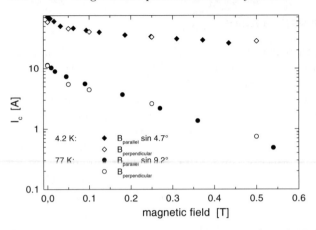

Fig. 11.5. Reduced representation of the magnetic-field dependence of j_c

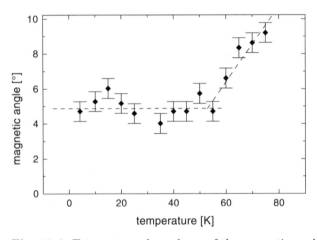

Fig. 11.6. Temperature dependence of the magnetic angle α

- As is well known from magneto-optical imaging (MOI) experiments [216], the current in Bi-2223 tapes flows mainly at the silver–BSCCO interface. This is assumed to be a result of the higher texture in this area. The coupling of individual grains increases with decreasing temperature; thus more and more grains from the middle part of the superconducting core also take part in the overall current path. These grains are less well aligned; thus the average misalignment angle should increase with decreasing temperature. As this is not the case, this explanation fails.
- As already mentioned above, the j_c anisotropy has intrinsic reasons and can be explained within a model of weakly coupled superconducting layers given by Lawrence and Doniach [156]. The increase in anisotropy with increasing temperature thus gives evidence for a reduction of the coupling strength between the superconducting layers. With increasing temperature

the pancake vortices within the Cu–O_2 planes decouple owing to thermal activation. Pancake vortices can move freely within a plane and are no longer coupled between planes. This leads to a reduced line tension of the vortices and thus to a more two-dimensional behavior of the flux lines. Consequently the pinning becomes more inefficient and thus the critical current density decreases.

The temperature at which the decoupling happens agrees well with the transition temperature found from the scaling analysis. This implies again that this transition is more likely to be explained by thermally activated flux creep than by a vortex phase transition. This interpretation does not exclude the existence of a phase transition, especially as in thin epitaxial BSCCO films evidence for such a transition has been found [263]. It just means that current–voltage characteristics are not sufficient for the analysis of phase transitions. Real thermodynamic quantities like specific heat have to be analyzed in order to obtain exact information.

11.4 Flux Creep Resistivity

The vortex glass model and the Kim–Anderson model of thermally activated flux creep differ in the assumption that in a vortex glass the resistivity should truly approach zero at low temperatures, while in the latter there is always a nonzero resistivity at nonzero temperatures. The flux creep resistivity ρ_{flux} was determined from the I–V curves at currents well below I_c and is depicted logarithmically against $1/T$ in Fig. 11.7 for various magnetic fields. This figure shows Arrhenius behavior for the flux creep resistivity, i.e.

$$\rho_{\text{flux}} \propto \exp(U/k_{\text{B}}T) , \qquad (11.10)$$

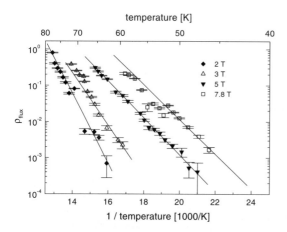

Fig. 11.7. Arrhenius plot of the flux creep resistivity ρ_{flux} for different magnetic fields

where U is the activation energy according to the Kim–Anderson model as discussed in Sect. 4.3. The slope of these curves is a measure of the activation energy. It becomes lower with increasing magnetic field as the potential wells in which the vortices are pinned become shallower. The behavior of the flux creep resistivity is a further hint that the degradation of the critical current density of BSCCO tapes in external magnetic fields with increasing temperature is caused by thermally activated flux creep.

11.5 Irreversibility Field

The irreversibility field H_{irr} is an important quantity for applications as it denotes the limit above which the magnetization of a hard superconductor becomes reversible owing to flux motion. John Anderson and coworkers could show that it is also possible to determine H_{irr} from I–V curves instead from magnetic hysteresis measurements [8]. For this purpose the I–V curves are plotted at constant temperature at different fields and the irreversibility field is then determined from the change in curvature.

The experiments described above have also been performed for Bi-2212 wires. A scaling behavior of the current–voltage characteristics could also be observed in this case [270]. However, as the degree of texture in round wires is very low it does not make sense to draw any conclusions from this evaluation. Nevertheless it is possible to depict the irreversibility field for both conductor types. In Fig. 11.8 these results are shown in comparison with those of Kiuchi et al. [126].

This irreversibility line reveals clearly the present limitations of the application of BSCCO conductors. For Bi-2223 tapes in magnetic fields normal to the tape plane the irreversible region lies below 50 K for magnetic fields below 2 T, and for Bi-2212 it lies below 30 K for the same field region. In order to

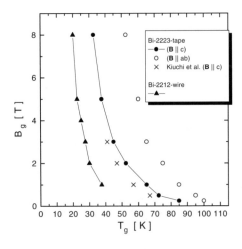

Fig. 11.8. Irreversibility field of Bi-2223 tapes with different orientations of the external magnetic field and of Bi-2212 wires, compared with results of [126]

produce higher magnetic fields, devices made of these superconductors have to be operated at temperatures below 40 K or 20 K, respectively.

12. Superconducting Magnetic Levitation

12.1 Levitation Force Experiments on Various Superconductors

Some possible applications of ceramic superconductors in magnetic bearings have been addressed in the introduction. One important quantity in these applications is the levitation force which is exerted on a permanent magnet by a superconducting ceramic sample. The basic form of levitation is caused by the Meissner–Ochsenfeld effect. Owing to the complete expulsion of magnetic flux out of a type I superconductor a permanent magnet is levitated above it. This effect is used as an experimental proof of superconductivity in HTSCs. However, it can only be demonstrated in type I superconductors or in type II superconductors below H_{c1}. Melt-processed YBCO ceramics, which are used for magnetic-bearing application, are extreme type II superconductors. That means that magnetic flux can enter the sample as soon as $H_{\text{ext}} > H_{c1}$. Good-quality melt-grown ceramics are hard superconductors; thus flux is pinned. This leads to a further physical effect, *magnetic suspension*, when a permanent magnet is suspended below a superconductor. An important consequence is: *The Meissner–Ochsenfeld effect cannot be demonstrated with a high-quality melt-grown $YBa_2Cu_3O_{7-\delta}$ sample above H_{c1}!* The physical reason for this will be discussed below.

12.1.1 Experimental Setup

A three-dimensional (3-D) positioning system was set up for the levitation force experiments. A schematic view of this setup is given in Fig. 12.1. The force sensor can be moved in the y and z directions while the sample is moved in the x direction. A permanent magnet attached to the force sensor can thus be moved along the surface of the superconductor and the distance between the superconductor and magnet can be varied. The motion is computer-controlled. Dc servo motors are used to control the three axes. A control circuit provides exact information about the actual position with an accuracy of 0.5 μm. The reproducibility is better than 5 μm in all three directions. The sample is placed on a copper rod within a Styrofoam container, which is filled with liquid nitrogen.

140   12. Superconducting Magnetic Levitation

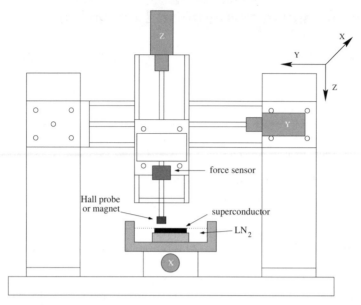

Fig. 12.1. Schematic setup of the three-dimensional positioning system

The force between a permanent magnet and a superconductor is measured by an axial strain gauge force sensor. The strain gauge translates the force into an electrical signal. The resistance change, which is proportional to the force, is measured by a Wheatstone bridge. The maximum recordable force of the chosen sensor is 10 N; the linearity as quoted by the provider is about 0.2%. The output signal of the Wheatstone bridge is monitored by a computer. In order to obtain reasonable thermal and magnetic insulation of the sensor, the test magnet was mounted on a PVC rod at a certain distance from the sensor.

Permanent magnets of different material, geometry and remanent field were used in the experiments. The test magnets, with their relevant parameters, are listed in Table 12.1. The remanent field was experimentally determined using a Hall sensor.

The setup is completely computer-controlled. A view of the apparatus is shown in Fig. 12.2. A more detailed discussion, especially of the calibration procedures, can be found in [147, 176].

12.1.2 Experimental Parameters

As in all other magnetic measurements, geometrical parameters play a crucial role in levitation force experiments. The force between the magnet and the superconductor depends not only on sample quality but also on the following parameters:

12.1 Levitation Force Experiments on Various Superconductors

Table 12.1. The different test magnets used for the levitation force experiments

Type	Material	h_a (mm)	$2\,r_a$ (mm)	B_r (mT) after [107]	B_r (mT) measured	$B_z(0)$ (mT) calculated
NE24	NdFeB	4	2	1100–1250	1072	520
BA519	SmCo	5	10	550–590	581	205
BF105	BaFe	5	10	170	172	60
NE105	NdFeB	5	10	1100–1250	1147	406
NE257	NdFeB	7	25	1100–1250	1154	282
DE105	SmCo	5	10	850–930	916	324
DE205	SmCo	5	20	850–930	1011	226
BF105	BaFe	5	10	245	247	87
NE103	NdFeB	3	10	1100–1250		296
NE110	NdFeB	10	10	1100–1250		514

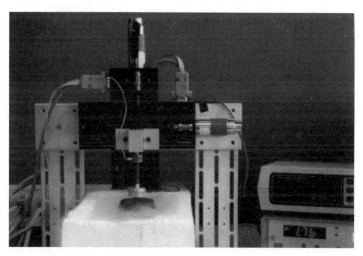

Fig. 12.2. Experimental setup for the measurement of spatial force and magnetic flux distributions

- sample size
- magnet size
- minimum distance between superconductor and magnet during experiments
- remanent magnetic field of the test magnet.

A real comparison of experimental results of levitation force measurements between different laboratories is only possible if an identical geometry of the setup is used or the setup is exactly known. The latter, however, only makes sense if the dependence of the levitation force on the geometry is known.

Influence of Geometry

Figure 12.3 shows the geometrical parameters of a simple levitation force experiment. The diameter and height of the magnet and superconductor determine the geometrical arrangement of the experiment. In order to test the parameter set D_S, H_S, D_M, H_M the sizes of the magnet and superconductor have to be varied. In the case of the magnets, commercially available materials with different diameters and heights were used. However, the sample size can not be varied as easily while ensuring equal quality. Different batches of ceramics often do not have exactly the same properties. Therefore the experiments concentrated on variation of the magnet dimensions.

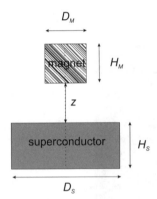

Fig. 12.3. Geometrical arrangement and quantities of a levitation force experiment

Figure 12.4 shows the force between a superconductor and test magnets of different lengths. The distance z between the superconductor and test magnet serves as a parameter. The curves show an asymptotic behavior of the force as a function of magnet length. This has to be expected as the surface field of a permanent magnet approaches the limit of infinite length with increasing height.

As a second parameter, the magnet diameter was varied between 4 mm and 30 mm while the length was kept constant at 5 mm. The results are depicted in Fig. 12.5 together with calculated results by Tsuchimoto et al. [257]. It can be seen that, with increasing diameter of the magnet, the force increases rapidly until it reaches its maximum, when the diameter of the magnet is equal to the distance between the magnet and superconductor. A further increase of magnet diameter leads to a reduction of the levitation force. The reason is that the remanent field of a cylindrical magnet depends on its aspect ratio H_M/D_M.

As already mentioned, the variation of the sample dimensions is not quite as easy as the variation of magnet size. Therefore the influence of the geometry of the superconductor was calculated using a diamagnetic model. In this model the superconductor is considered as an ideal diamagnet. Its diameter is

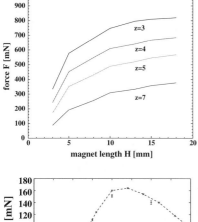

Fig. 12.4. Force between magnet and superconductor for different distances z as a function of magnet length

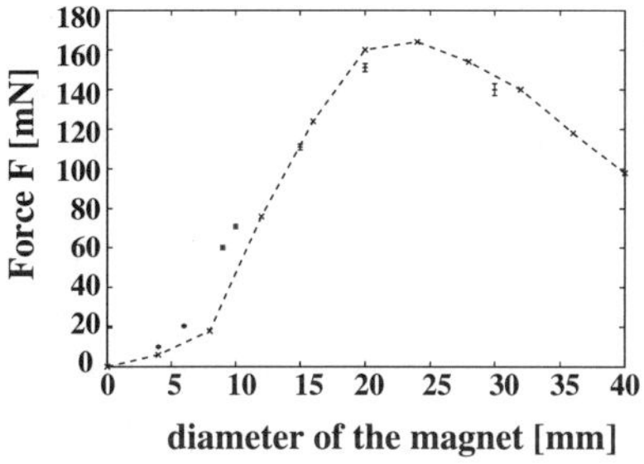

Fig. 12.5. Force at a constant distance of 2.5 mm between magnet and superconductor as a function of magnet diameter. The *dashed line* represents a calculated curve by Tsuchimoto and coworkers [257]

chosen as at least twice the diameter of the magnet; thus the sample appears to be infinitely large compared with the test magnet. Then the arrangement depicted in Fig. 12.3 is that of a magnet above a semi-infinite superconducting plane. In analogy with the electrostatic problem of mirror charges, the setup can be treated as two magnets with opposite magnetization and twice the distance between them. The situation is illustrated in Fig. 12.6.

The field distribution of this geometry can be calculated analytically and numerically. From the field distribution, the force can be determined using the Maxwell stress tensor. The results from such simulations were compared with those of Tsuchimoto, obtained with a more complex model. Figure 12.7 shows the calculated force between the superconductor and magnet as a function of

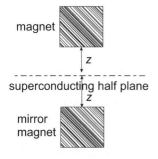

Fig. 12.6. Diamagnetic analogue of the mirror charge problem. A magnet at distance z above a superconductor can be considered as two magnets with opposite magnetization and twice the distance between them

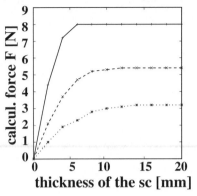 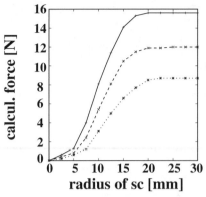

Fig. 12.7. Calculated force between magnet and superconductor as a function of superconductor thickness

Fig. 12.8. Calculated force between magnet and superconductor as a function of the radius of the superconductor

superconductor thickness. The parameter is the quality of the superconducting sample. The radius and length of the magnet (R_M, H_M), and the radius of the superconductor (R_S) were kept constant at 10 mm. The distance z between the superconductor and magnet was chosen to be 3 mm. The force increases with increasing height of the sample and saturates at $H_M = H_S$. That means that an infinitely long sample perfectly screens the field of the magnet.

Figure 12.8 shows a similar calculation, where the radius of the superconductor was varied instead. The other conditions were the same. The same behavior is observed: the levitation force increases with increasing sample diameter and reaches a saturation value at $R_S = 2 \times R_M$. The reason is that the remanent field of the magnet approaches that of an infinitely long cylinder.

In the following experiments the sample-to-magnet geometry was chosen as $D_S : D_M = 2.2 : 1$. Thus the influence of the geometry, according to Fig. 12.8, was at its saturation value and a ten percent variation of sample diameter would not have an influence on the measured force.

Remanent Field of the Test Magnet

In order to determine the influence of the remanent field of the test magnet on the measured force, experiments were performed on one sample with various magnets from those listed in Table 12.1. The levitation force was measured as a function of the distance z between the test magnet and sample. The results are shown in Fig. 12.9.

The results for the Betaflex and BaFe magnets (denoted BA and BF in Table 12.1) are very similar within experimental accuracy. The results for the Delta and Neodelta magnets (denoted DE and NE in Table 12.1) are also similar to each other. The weaker Betaflex and BaFe magnets exhibit a

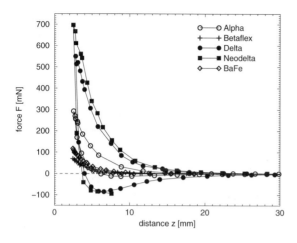

Fig. 12.9. Force–distance characteristics of different magnets with superconducting sample P-SO-13

smaller hysteresis, though. The stronger Delta and Neodelta magnets have pronounced hysteretic behavior and greater repulsive and attractive forces. This is evidence that the field strengths of the Betaflex and BaFe magnets are not sufficient to fully penetrate the superconductor, while the Delta and Neodelta magnets are strong enough for full penetration.

Figure 12.10 shows experimental results of levitation force measurements at a distance of 2.5 cm as a function of the remanent field of the test magnet. The dependence is linear up to the maximum field used, as would be expected

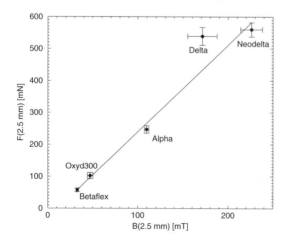

Fig. 12.10. Maximum force as a function of the remanent field of various test magnets

from the Lorentz force (4.16). Thus it can be concluded that the Bean critical-state model is a good approximation for melt-grown $YBa_2Cu_3O_{7-\delta}$ samples.

Maximum Possible Levitation Force

It is possible to normalize levitation force results with known geometry. The levitation force is calculated using the mirror charge method. As a result the force between two permanent magnets with opposite charge is obtained, corresponding to the force of a permanent magnet above a superconducting half plane. An expression for this force can be deduced from the London theory and the Maxwell equations [176]. The maximum achievable force $F^{max}(x)$ corresponds to the force between these two permanent magnets with a distance between them of $2z$. Riise et al. [232] give an expression for this force

$$F_z(z) = -MB_r f(z). \tag{12.1}$$

In this expression the function $f(x)$ depends only on geometry. It can be found in [176] and is used to calibrate the experiment. This maximum obtainable force is independent of sample quality and is determined only by the geometry and the properties of the test magnet. That means that there is an upper limit on the levitation force that depends on the field strength of the permanent magnet used.

12.1.3 Comparison of Various Superconductors

Taking into account the above considerations, various types of superconductors were examined. The most relevant results of some typical samples will be presented below. Three of the four samples were prepared in our group; the fourth sample was prepared by the Institut für Physikalische Hochtechnologie (IPHT) in Jena.

- Sample SB110 is a c axis-textured epitaxial $YBa_2Cu_3O_{7-\delta}$ thin film [164]. It differs from the other three samples owing to its nearly two-dimensional geometry. As a quasi-single-crystal superconductor, it has a very high critical current density as compared with the melt-grown samples. However, for an $YBa_2Cu_3O_{7-\delta}$ film from our laboratory its j_c and T_c are untypically low.
- The sintered pellet Sinter-1 is a very-fine-grained polycrystalline, untextured sample of pressed and sintered YBCO powder [160, 161, 219].
- The melt-grown sample S00D-3 consists of several well-textured grains with amorphous material in between. As a precursor, $YBa_2Cu_3O_{7-\delta}$ in stoichiometric composition with 10% silver (Ag_2O) addition was used [160, 161, 219]. The sample was prepared using a modified LPP process [147, 176]

12.1 Levitation Force Experiments on Various Superconductors

- Sample JC284 is a melt-grown ceramic like S00D-3 but of better quality. It was prepared at the IPHT Jena. As a precursor $YBa_2Cu_3O_{7-\delta}$ powder with 20% Y_2O_3 and one weight percent platinum was used [63, 64].

In Table 12.2 the relevant sample parameters sample thickness H_S, radius R_S, density ρ, critical current density j_c at 77 K and transition temperature T_c are listed.

Table 12.2. Parameters of selected samples

Sample	H_S (mm)	R_S (mm)	ρ (g/cm^3)	j_c(77 K) (A/cm^2)	T_c (K)	Reference
SB110	5×10^{-4}	25.4	–	10^6	90	[23]
Sinter-1	4.8	21.2	5.9	10^2 (intergranular) 10^4 (intragranular)	92	[146]
S00D-3	6.8	21	5.8	10^3–10^4	92	[161]
JC284	9.4	29	5.7	10^3–10^4	92	[63, 64]

Zero-Field-Cooled Experiments

The levitation force as a function of distance between test magnet and superconductor was measured at 77 K after cooling in zero field and in an external magnetic field. The zero-field-cooled (zfc) results are shown in Fig. 12.11. The force–distance characteristics of all four samples are depicted in this plot. In the inset the results of the sintered sample and the thin film are drawn on a finer scale as their forces are much lower than those of the bulk samples.

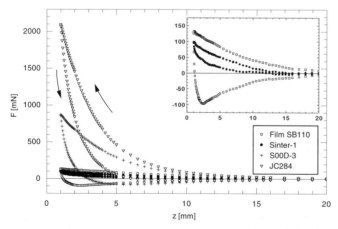

Fig. 12.11. Force–distance characteristics of four typical samples after zero-field cooling

The sintered sample exhibits only very small hysteresis and no attractive force upon removal of the magnet. The reason is that in this fine-grained sample magnetic flux can easily enter and leave the superconductor. There is nearly no flux pinning. The epitaxial thin film has a low levitation force on an absolute scale but exhibits pronounced hysteretic behavior. The attractive force is nearly as large as the repulsive force. This is a signature that the film is completely penetrated by magnetic flux. The flux lines are well pinned so a counterforce has to act to withdraw them from the sample. The bulk samples are of different quality and thus exhibit a different levitation force. The main difference, however, is that, as in the sintered sample, there is no attractive force in the case of the better sample JC284. This interesting fact can be understood phenomenologically. If a hard superconductor is zero-field cooled and a magnetic field is applied after it has gone into the superconducting state, flux cannot easily enter the sample as it is efficiently pinned already in the outer regions of the sample. According to Brandt [20] an ideally hard superconductor cooled in zero field behaves like a type I superconductor in the Meissner state. Consequently magnetic flux cannot easily penetrate a high-quality melt-grown sample after it has been cooled without external magnetic field.

Field-Cooled Experiments

During field cooling (fc) a test magnet was kept at a constant distance of 1 mm above the sample. The results are shown in Fig. 12.12. As in Fig. 12.11 the results of the sintered sample and the thin film are shown enlarged in the inset.

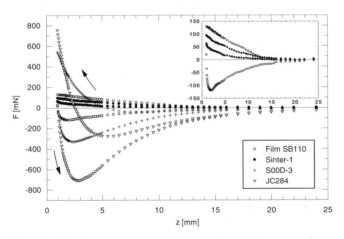

Fig. 12.12. Force–distance characteristics of four typical samples after cooling in an external field

It can be seen that the force–distance characteristics of the latter two samples are not very different from the zfc case. The bulk samples, however, exhibit strong hysteretic behavior. The forces are attractive upon approaching and upon removal as well. This is a consequence of the pinned magnetic flux which has entered the superconductor during cooling.

12.1.4 Spatial Distribution of Levitation Force

In the experiments presented up to now the levitation force was measured as an integral quantity. For a more thorough characterization, however, it is useful to determine the force distribution over the whole sample surface. For bearing applications homogeneous sample properties are desirable. The spatially resolved measurements were performed using a small test magnet of only a few millimeters in diameter in order to obtain a reasonable lateral resolution. The test magnet was scanned over the sample surface using the 3-D scanning setup described above. The distance between the sample and magnet was kept constant between 0.55 mm and 1 mm. The samples were field-cooled and the force map was monitored at 77 K. In addition the magnetic flux distribution was recorded using a Hall sensor. The results of the two sets of experiments for the four selected samples are shown in Figs. 12.13 and 12.14 for comparison.

The thin film and sintered sample exhibit a quite homogeneous force distribution. Nevertheless the comparison with the magnetic-flux maps reveals that the flux is evenly distributed in the sintered pellet while it shows a pronounced Bean cone in the epitaxial thin film. The cone is slightly asymmetric, which is a consequence of the preparation process (nonhomogeneously heated substrate). The melt-grown samples show an inhomogeneous force distribution, which is correlated with an inhomogeneous distribution, of magnetic flux as well. The bulk samples exhibit several Bean cones, which is a signature of the multidomain structure of these samples.

By comparing the applied external field with the trapped magnetic flux a rough estimate of the pinning properties can be obtained. The quantity $\epsilon_B = B_{z(\text{sample})}^{\max}/B_{z(\text{magnet})}$ relates the maximum flux density measured above the sample to the flux density of the magnet used to magnetize it. This pinning efficiency ϵ_B is compared with the levitation force in Table 12.3. This

Table 12.3. Maximum values of force and remanent flux from the force and field maps, and efficiency ϵ_B

Sample	$F_{\max}^{\text{scan}}$ (mN)	$B_{z,\max}^{\text{scan}}$ (mT)	ϵ_B (%)
SB110	24	11.3	4.3
Sinter-1	16	0.8	0.3
S00D-3	69	31.0	11.9
JC284	91	72.5	27.9

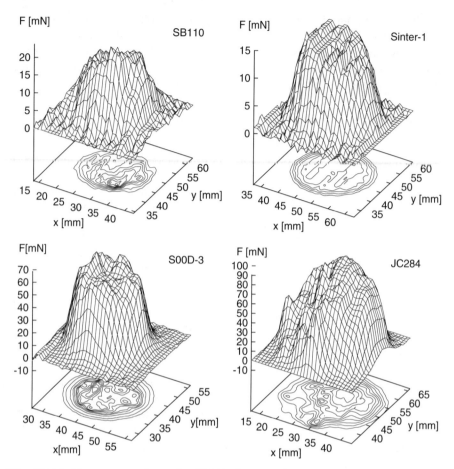

Fig. 12.13. Force maps of samples SB110, Sinter-1, S00D-3 and JC284

comparison shows that the levitation force is correlated with the pinning properties of the sample, as the one with the highest trapped flux also shows the greatest levitation force. Comparing the two spatially resolved measurements of levitation force and trapped flux, it is clear that the remanent-flux distribution contains more interesting information about the sample quality than the levitation force distribution.

Single-domain samples have been prepared by several groups around the world. These melt-grown $YBa_2Cu_3O_{7-\delta}$ ceramics consequently exhibit one single, highly symmetric Bean cone in the measured flux distribution. The focus of this work, however, was to obtain more insight into the basic physics of superconducting levitation; thus sample optimization was not performed. Moreover, several models will be proposed in the following which may offer some deeper understanding of the physics of superconducting levitation.

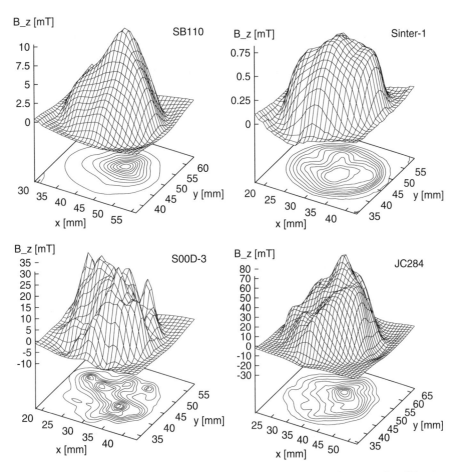

Fig. 12.14. Remanent flux distribution after field cooling for samples SB110, Sinter-1, S00D-3 and JC284

12.2 Understanding Force–Distance Hysteresis

There are two physical origins for the force that a superconductor exerts on a permanent magnet. Below H_{c1} the Meissner effect induces circular currents within the superconductor when it is placed in an external magnetic field. This leads to a repulsive force between the superconductor and magnet. Above H_{c1} magnetic flux can enter the superconductor as vortices. The Lorentz force acts on these flux lines. In the case of a hard superconductor the pinning force counteracts the Lorentz force. The approach and removal of a permanent magnet are connected with flux motion, which is hindered by the pinning force. Thus the higher the pinning force, the higher the force necessary for a test magnet to approach. Upon removal of the magnet, flux lines have to leave the superconductor. If the pinning force is very high this causes an attractive force and the sample can be suspended underneath the magnet.

12.2.1 Interpretation Within the Bean Critical-State Model

We shall apply the Bean critical-state model, which was introduced in Sect. 4.4, for the interpretation of levitation force experiments. In an external magnetic field with induction $B_{\text{ext}} = \mu_0 H_{\text{ext}}$, a force

$$\boldsymbol{F} = \int_V (\boldsymbol{M} \cdot \boldsymbol{\nabla}) \boldsymbol{B}_{\text{ext}} \, \mathrm{d}V \qquad (12.2)$$

acts on the flux lines. If the magnetization is known and the magnetic field is oriented parallel to the z axis in Fig. 12.3, (12.2) can be reduced to

$$F = -F_{\text{PM}} = VM \frac{\mathrm{d}B_{\text{ext}}}{\mathrm{d}z} \, . \qquad (12.3)$$

Brandt [20] deduces the magnetization of an elliptical sample with strong pinning from the Bean model. Taking into account the demagnetization effects in the given geometry (Fig. 12.3), the magnetization curve of Fig. 12.15 can be drawn. From this magnetization curve a force–distance characteristic can be calculated using the Lorentz force (4.16) and assuming that the field of the permanent magnet may be idealized by a dipole field. This calculation leads to Fig. 12.16. The different sections of this curve can be explained as follows. Upon the first approach of a magnet towards a superconductor the magnetization $M(B_{\text{ext}})$ of the sample follows a virgin curve (broken line). According to (12.3), from $\mathrm{d}B_{\text{ext}}/\mathrm{d}z > 0$ and $M < 0$ a repulsive force $F_{\text{PM}} > 0$ follows.

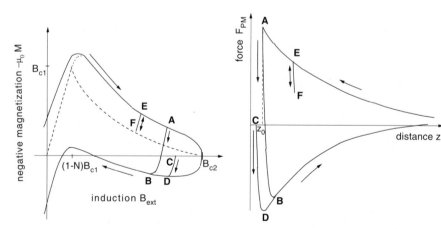

Fig. 12.15. Magnetization hysteresis of a hard superconductor with demagnetization effects (after [20])

Fig. 12.16. Force–distance characteristic from the magnetization curve shown in Fig. 12.15 (after [20])

When the magnet has reached a minimum distance z_0 it is withdrawn from the superconductor and consequently the magnetization follows an inner loop along A–B, changing sign from positive to negative after a small decrease of the external magnetic field. This implies an attractive force $F < 0$ between the superconductor and magnet (suspension). If the superconductor is field cooled (point C in Fig. 12.15), magnetic flux is trapped in the sample below the transition temperature. Owing to its pinning forces the superconductor tends to keep the vortices in place, which leads to a positive magnetization $M > 0$ and thus to an attractive force. Upon removal of the magnet, the magnetization and force follow the line C–D.

If the magnet performs a minor loop by moving a small distance opposite to the original direction, the magnetization follows an inner loop of the maximum hysteresis loop, corresponding to line E–F. This inner loop initially has the same slope as the virgin curve.

If the idealized force–distance curve in Fig. 12.16 is compared with the experimental results of Figs. 12.11 and 12.12 it can be seen that only in the case of the thin film does the levitation force as a function of distance have a similar shape. The other results do not resemble the idealized curve. The reason is that the description of a superconductor in an external magnetic field within a magnetization model is not always appropriate. Originally meant for the description of matter in magnetic fields, magnetization is assumed to be a superposition of microscopic screening currents. In the case of a superconductor, however, this picture can lead to nonphysical effects. For example, if a superconducting ring is field cooled into the superconducting state flux is expelled from the sample owing to the Meissner effect. As a consequence magnetic flux is present inside the ring. When the external field is now reduced to zero this flux remains in the interior of the ring as, owing to the persistent currents within the superconductor, it cannot leave through the sample. Therefore in the interior $B = \mu_0(H + M) > 0$. With $H = 0$ it follows that $M > 0$ and thus the interior of the ring has a nonzero magnetization despite the fact that there is no matter and thus no screening currents [225].

Hence, in the following model, first proposed by Portis [225], the levitation force will be deduced from the interaction of the screening currents within the superconductor with the external magnetic field. The force exerted on a current carrying-conductor by an external magnetic field B_{ext}/μ_0 is

$$\boldsymbol{F} = \int_V \boldsymbol{j} \times \boldsymbol{B}_{\text{ext}} \, \text{dV} \,. \tag{12.4}$$

The current distribution $\boldsymbol{j}$ inside the superconductor can be expressed as an effective surface current per unit length I_0. Its magnitude as a function of the external magnetic field can be deduced from the Bean model under the following assumptions:

- The sample is a rod with thickness r_g.
- The external magnetic field is homogeneous and parallel to the cylinder axis.

The effective surface current I_0 increases linearly with increasing flux density B_s at the surface of the rod until it reaches its maximum value $I_0 = j_c r_g$ at $B_s = B^*/2 = \mu_0 j_c r_g$, where B^* is the full-penetration field. The flux density at the surface of the rod is a superposition of the flux density B_{ext} caused by the external field and that caused by the surface currents I_s:

$$B_s = B_{\text{ext}} + \mu_0 I_s/2 = B_{\text{ext}} + (I_s/I_0) B^*/2. \tag{12.5}$$

From this equation the effective screening current as a function of the external field $I_s(B_{\text{ext}})$ has been estimated and is depicted in Fig. 12.17.

Using the Lorentz equation (4.16), the force per unit surface area is

$$F/S = I_s B_{\text{ext}}. \tag{12.6}$$

In the region of the virgin curve, i.e. for $0 < B_{\text{ext}} < B^*/2$, $I_s = I_0 B_{\text{ext}}/(B^*/2)$ is valid. The force density increases quadratically with external field:

$$\frac{F}{S} = 2 \frac{I_0}{B^*} B_{\text{ext}}^2. \tag{12.7}$$

For flux densities $B_{\text{ext}} \geq B^*/2$ the surface current $I_s = I_0$ is constant and therefore the force per unit surface area increases linearly with magnetic field:

$$\frac{F}{S} = I_0 B_{\text{ext}}. \tag{12.8}$$

When the flux density is reduced from $B_{\text{ext}}^{\max}$ the surface current decreases linearly from $+I_0$ to $-I_0$ and the force density follows a parabola:

$$\frac{F}{S} = 2\frac{I_0}{B^*}\left[B_{\text{ext}}^2 + \left(\frac{B^*}{2} - B_{\text{ext}}^{\max}\right) B_{\text{ext}}\right]. \tag{12.9}$$

If a complete cycle of the external flux density is performed, the characteristic "butterfly" form of Fig. 12.18 is obtained.

In order to compare this model with the experimental results, the data from Fig. 12.11 has been depicted as a function of the external flux density B_{ext}, which is obtained from the flux density of the test magnet and the corresponding distance between the magnet and sample. The levitation force as a function of external magnetic field is shown in Fig. 12.19 for all four samples. These curves exhibit the "butterfly" form; however, there is no sharp edge at $B_{\text{ext}} = B^*$ or $B_{\text{ext}} = B^*/2$.

Up to now the force density has been considered. To obtain the absolute levitation force the sample geometry has to be taken into account. The force is calculated from the force density by integrating over the whole surface of the superconductor:

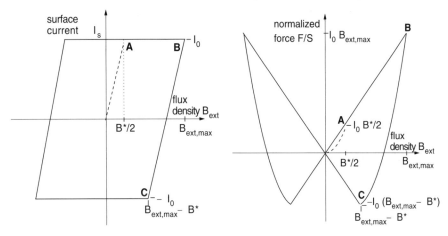

Fig. 12.17. Screening current I_s as a function of B_{ext} (after Portis [225])

Fig. 12.18. Normalized force F/S as a function of B_{ext} (after Portis [225])

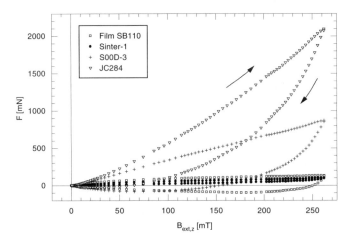

Fig. 12.19. Levitation force as a function of the external flux density B_{ext} for all four selected samples

$$F = 2\pi \int_0^R \mathrm{dr} I(r) H(r). \tag{12.10}$$

A detailed discussion of the numerical solution of this integral is given in [147]. This integration means that large radii contribute more to the overall force, thus leading to a rounding of the shape of the force–distance curves. In Fig. 12.20 the simulation is compared with the experimental results for the melt-grown sample JC284. The qualitative agreement is quite good, so it can be concluded that the Bean model provides a good means to describe the force–distance characteristics of melt-grown ceramics. An even better fitting of the experimental results is obtained by considering volume currents as

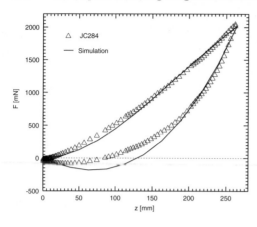

Fig. 12.20. Experimental and computed results for the levitation force as a function of external flux density for a melt-grown sample

well. This complex three-dimensional model has been calculated by Tsuchimoto and coworkers [257]. The results for the epitaxial thin films, however, cannot be fitted as well. An important restriction of the Bean model, the field-independent critical current density, has to be omitted to explain the film results.

12.2.2 Magnetic-Field-Dependent Critical Current Density

Anderson and Kim [10] proposed a modification of the Bean model by introducing a field-dependent critical current density of the form

$$j(B) = \frac{2j_0}{1 + (B/B_0)^\beta} \,. \tag{12.11}$$

In this general expression j_0, B_0 and β are fit parameters. The exponent β denotes which of the two models is valid:

- $\beta = 0$: Bean model
- $\beta = 1$: Kim–Anderson model.

Calculations from the model using the data from the thin-film samples yields a best-fit value of $\beta = 0.8$. This does not quite correspond to the Kim–Anderson model. The experimental results and simulation are shown in Fig. 12.21.

It is clear that neither the Bean model nor the Kim–Anderson model is exactly valid. The best fit is obtained with a modified Kim–Anderson model. This can be attributed to the fact that the thin film does not fulfill certain assumptions made by both models. The film thickness is about 300 nm and therefore of the order of magnitude of the London penetration depth. Hence it is not a bulk sample, as assumed by Bean. Furthermore the high aspect ratio of the sample (diameter/thickness $= 3 \times 10^4$) leads to great demagnetization effects with $N > 1000$. This may cause even the earth's magnetic field to

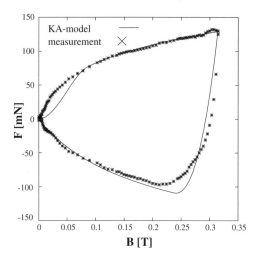

Fig. 12.21. Measured and calculated force on an epitaxial thin film as a function of the axial flux density

exceed the lower critical field H_{c1} at the borders of the sample. Consequently flux enters the superconductor, and the quadratic increase of the force at low fields, which is due to the Meissner effect, cannot be observed in the thin-film samples.

The results of the levitation force experiments on the sintered pellets are better described by the diamagnetic model. As these are granular samples there are no macroscopic screening currents present. Each micrograin has a microscopic screening current circulating around it instead. These microscopic currents support the use of the diamagnetic model for the description of these results. The field-dependent representation of the levitation force measurements of the sintered samples thus exhibits the diamagnetic analogue of a ferromagnetic hysteresis.

In conclusion it can be said that the zfc results of the four selected samples cannot be described concisely with one model. While the sintered sample is best described by the diamagnetic model, the melt-grown samples are well approximated by the Bean critical-state model. The thin film, however, can only be described by a modified Kim–Amderson model that considers magnetic-field-dependent critical current densities. Nevertheless, the results of all four samples can only be approximated qualitatively by the models used. A quantitative analysis of physical quantities such as j_c, B^* and grain size is not possible.

12.3 Interpretation of Vertical-Stiffness Experiments

Besides the levitation force, there is a further physical quantity that is interesting for magnetic-bearing applications. The magnetic stiffness η characterizes the stiffness of a magnetic bearing against vertical or lateral movement of the load. But it also contains some interesting basic physical information. Therefore we shall differentiate between η as a bearing parameter and as a physically interesting quantity.

12.3.1 Vertical Magnetic Stiffness as a Bearing Parameter

In our experiments only the vertical magnetic stiffness was measured. The measurement was made within a force–distance experiment by performing small reversible variations of the distance within the force–distance hysteresis – so-called *minor loops*. An example of such an experiment is shown in Fig. 12.22. The inset shows two minor loops and the linear regression used to calculate η.

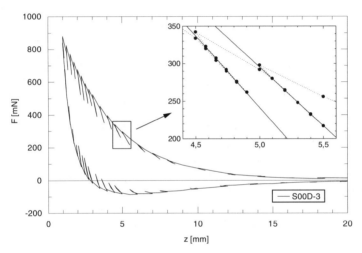

Fig. 12.22. Measurement of minor loops during a levitation force experiment; two loops with linear regression for the determination of η are shown in the *inset*

If the displacement of the minor loops is in the reversible regime, $\eta = \mathrm{d}F/\mathrm{d}z$ is valid. Moon [198] gives an empirical correlation between stiffness and levitation force:

$$\eta \propto F^{\beta}. \tag{12.12}$$

A comparison of this empirical formula and the results for the four standard samples is shown in Fig. 12.23. For small forces ($F < 100\,\mathrm{mN}$ for the film and sintered samples, $F < 300\,\mathrm{mN}$ for melt-grown samples) the exponent β is constant to a first approximation. At higher forces β increases, with the exception of the sintered sample. The strongest increase is exhibited by the thin film. A linear regression in the region of small forces yields $\eta = F^{2.8}$ for the thin-film and sintered samples, and $\eta = F^{1.2}$ for the melt-grown samples.

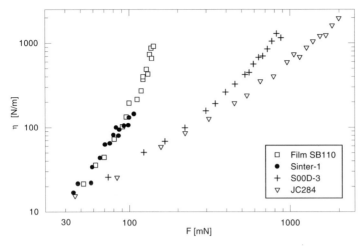

Fig. 12.23. Vertical stiffness η as a function of levitation force for the four selected samples

The exponent of the melt-grown samples lies within the region of 1.0–1.6 as given by Moon [198]. This is not valid for the epitaxial thin film and the granular sample. In particular, the increase of vertical magnetic stiffness is surprisingly large for the thin film.

For magnetic-bearing applications, the properties of superconductors have to be compared with conventional magnet or air bearings. The samples which are available now, for example at IPHT Jena or FZ Karlsruhe, have much better quality than the ones discussed here. Nevertheless the stiffness of air bearings is not reached. A small air bearing has a vertical stiffness of 5 N/µm over an area of one cm^2 [141]. The highest stiffness values measured in our experiments are 1 mN/µm over the same area.

12.3.2 Stiffness and Labusch Parameter

Despite the very low vertical magnetic stiffness as compared with practical bearings, the measurement of η as well as that of the levitation force offers an opportunity to obtain more insight into the basic physical phenomena. As already mentioned, the physical origin of the force between a superconductor and a permanent magnet is the magnetic flux trapped in the Shubnikov phase of the superconductor. Variations of the external magnetic field caused by movements of the magnet lead to flux motion. Thus a direct observation of the properties of flux lines in these superconductors is possible.

In particular, the vertical stiffness is connected directly with the pinning properties of the measured sample. To visualize this it is useful to plot the vertical magnetic stiffness as a function of the external magnetic field as was done in the case of the levitation force results. This is done in Fig. 12.24.

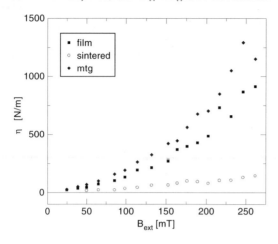

Fig. 12.24. Vertical stiffness as a function of the external flux density (mtg, melt-grown)

As in the η versus F plot a change in slope is observed with increasing field, except for the granular sample. At 120 mT the field dependence changes from $B^{1.5}$ to B^2.

A simple physical picture can be used to obtain a phenomenological interpretation of the stiffness results. On the basis of the basic discussion in Sect. 4.3, flux lines are considered to be bound in a harmonic pinning potential with the pinning energy U_p. The overall restoring force in a minor-loop experiment is

$$F_z = \eta z, \tag{12.13}$$

corresponding to the restoring force or spring constant of the vortex in the potential well. In equilibrium with an external magnetic field H_ext, the pinning force and Lorentz force are equal:

$$f_\mathrm{P} = \boldsymbol{j} \times \boldsymbol{B}. \tag{12.14}$$

This pinning-force density

$$f_\mathrm{P} = \mathrm{d}U_\mathrm{P}/\mathrm{d}z \tag{12.15}$$

is deduced from the pinning energy of one vortex in its pinning potential per unit volume. In the elastic regime the Lorentz force causes only reversible displacements of flux lines in their potential wells, as depicted schematically in Fig. 12.25.

If the critical current density is exceeded, flux lines can escape from their pinning potential and move through the superconductor, causing losses. The displacement becomes irreversible. From the measurement of the response of a superconductor to an external alternating magnetic field the elastic limit d_0 and the effective restoring force per unit volume $\alpha = \mathrm{d}f_\mathrm{p}/\mathrm{d}z$ can be obtained. As can be seen from Fig. 12.25 the elastic limit d_0 essentially corresponds

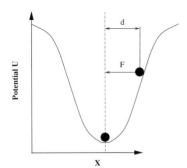

Fig. 12.25. Vortex in its harmonic pinning potential

to the lateral extension of the pinning potential, while α corresponds to its slope. The slope of a potential well is also expressed by its spring constant; thus there has to be a connection between η and α. The latter also called the Labusch parameter [151]. The quantity α is a force density per unit volume; thus η has to be related to a volume in order to connect it with α. According to [151, 187] it is therefore assumed that there is an effective volume V_{eff}, which is the correlation volume of the vortices. In a superconductor with multiple grains this has to be the grain size as the superconducting state is not very well defined over grain boundaries. Hence the pinning-force density

$$f_{\text{P}} = F_z/V_{\text{eff}} \tag{12.16}$$

and the Labusch parameter

$$\alpha = \mathrm{d}f_{\text{P}}/\mathrm{d}z = \eta/V_{\text{eff}} \tag{12.17}$$

are obtained from the measurement of the vertical stiffness η.

The Labusch parameter α as a function of external field is shown in Fig. 12.26 for three of the four standard samples. As the correlation volume

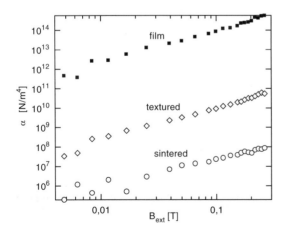

Fig. 12.26. Labusch parameter α as a function of the external flux density for three selected samples

a grain size of 2–5 mm was used for the melt-grown sample as this was the size of the crystallites observed in the optical microscope. This corresponds to an effective volume of $V_{\text{eff}} = 2 \times 10^{-8}\,\text{m}^3$. For the sintered sample and the epitaxial thin film the whole sample size was assumed as the correlation volume for different reasons (film: $V_{\text{eff}} = 2 \times 10^{-12}\,\text{m}^3$; sintered sample, $V_{\text{eff}} = 2 \times 10^{-6}\,\text{m}^3$). The sintered pellet is very fine-grained and does not show flux pinning, as deduced already from the levitation force experiments. Thus as an origin of the stiffness only the Meissner effect can be considered. The thin film is nearly monocrystalline so that the vortices are correlated over the entire volume. The results obtained for α are of the same order of magnitude as those measured with vibrating reeds [131, 287]. Our method is experimentally much simpler, though.

Nevertheless care has to be taken. Owing to the very simple picture of one vortex pinned in its individual potential well without nearest-neighbor interaction, the results can only be qualitative. That vortex–vortex interaction has in fact to be considered can be seen from the change of slope of η with increasing external field (Fig. 12.24). This is also reflected in the field dependence of α, as shown for the melt-grown sample in Fig. 12.27.

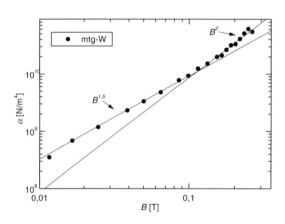

Fig. 12.27. Double-log representation of the field dependence of α for the melt-grown sample S00D-3. The *straight lines* are numerical fits to $\alpha = c\,B^n$

At about 120 mT the field dependence changes from $B^{1.5}$ to B^2. In this field regime the vortex spacing equals $a_0 = 0.2\,\mu\text{m}$. For larger spacings single-vortex pinning of one flux line in its potential well is dominant. At smaller spacings the repulsive interaction of vortices comes into play. Here the Labusch parameter α no longer denotes the stiffness of a single potential well, but is a measure of the elastic properties of the flux line lattice. As already discussed in Sect. 4.3, the compressional and shear moduli c_{11} and c_{44} of the flux line lattice are proportional the square of the external flux density:

$$c_{11} \propto c_{44} \propto \frac{B^2}{\mu_0}. \tag{12.18}$$

According to Brandt [21] the transition from single-vortex to collective pinning takes place at $B_{\text{ext}} > 2B_{c1}$. Thus the value for the lower critical field of YBa$_2$Cu$_3$O$_{7-\delta}$ extracted from the measurements presented here is 60 mT, which corresponds well to results from high-frequency experiments on thin films of YBCO [90]. It is in particular remarkable that the results for η do not depend on the history of the sample. This also gives evidence of the intrinsic physical origin of this quantity.

12.4 Experiments on Stacks of Epitaxial Thin Films

One of the most surprising results of the previous section was the relatively high levitation force and stiffness of the epitaxial thin films of YBa$_2$Cu$_3$O$_{7-\delta}$ as compared with their small volume. The question arises whether in a stack of such films the forces will add up. Therefore an experiment was performed in which the results of levitation force measurements of three single films were compared with those of a film stack and a melt-textured sample.

12.4.1 Force–Distance Characteristics

Three different thin films of YBa$_2$Cu$_3$O$_{7-\delta}$ with a diameter of 25 mm and thicknesses between 150 nm and 1500 nm were used. The properties of these films are summarized in Table 12.4. The films were prepared by high-oxygen-pressure dc sputtering. A more detailed explanation of the method can be found in [164]. The critical current density was measured inductively by detecting the third-harmonic susceptibility as described in [65].

Table 12.4. Properties of the measured YBa$_2$Cu$_3$O$_{7-\delta}$ films

Sample	d (nm)	T_c (K)	j_c(77K) (MA/cm^2)	F_0 (mN)
RW173	150	90.5	4.4	77.8
RT30	1500	89.7	0.29	63.4
MG40	500	88.9	3.0	91.5

Firstly, the force–distance characteristics of each film were measured separately. Then the films were glued on top of each other with OPTICLEAN^{TM} with their coated sides towards the magnet. The same measurement was subsequently performed with the film stack. All experiments were performed zero-field cooled. The single-film results are shown in Fig. 12.28. All three

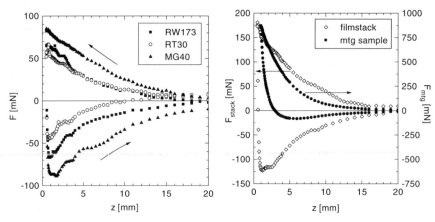

Fig. 12.28. Force–distance characteristics of three single films

Fig. 12.29. Force–distance characteristics of a stack of three films compared with a melt-textured sample

films exhibit a pronounced hysteresis. The attractive and repulsive forces are nearly equal, which implies that all films are fully magnetized.

The results from the film stack are shown in Fig. 12.29 together with those of a melt-grown sample. The levitation force of the film stack is one-fifth of that of the bulk sample. However, the hysteretic behavior deviates strongly. The attractive force of the stack is still nearly as large as the repulsive force, whereas the repulsive force of the bulk sample exceeds its attractive force by nearly a factor of ten. Nevertheless the small difference between the repulsive and attractive force of the film stack already implies that flux penetration is no longer perfect.

Figure 12.30 shows the approaching branch of the force distance–characteristics of the film stack compared with the sum of the single-film results. The stack results and the sum of the single-film results obviously differ. In part this can be explained by the increasing distance from the magnet of the films within the stack. The curve corrected for this distance variation is also depicted in Fig. 12.30. However, the overall levitation force of the film stack remains 15% lower than the sum of the single-film results. This can only be explained by screening of the external flux density by the upper films. A theoretical interpretation of our results was obtained by Tsuchimoto and coworkers [255, 256].

12.4.2 Magnetic Stiffness

Finally, the vertical magnetic stiffness of the film stack was measured. Again the results are depicted in comparison with results on a bulk sample, in Fig. 12.31. The stiffness of the stack is comparable to that of the bulk sample.

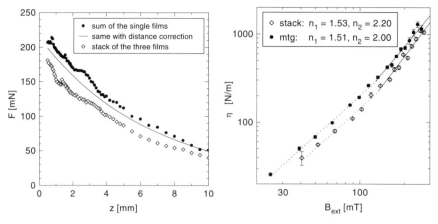

Fig. 12.30. Comparison of the results on the film stack with the sum of the single-film measurements with and without distance correction (see text)

Fig. 12.31. Vertical magnetic stiffness of the film stack as a function of the external magnetic flux density compared with the results from a melt-grown sample

Also, the same magnetic-field dependence as discussed above can be found in these results.

A simple estimation [177] shows that a stack of ten films of currently available quality in the geometry used would yield a levitation force of 1 N. Thick $YBa_2Cu_3O_{7-\delta}$ films as prepared in other laboratories [58] could provide similar values, according to this estimation.

13. Remanent Flux Distribution and Critical Current Density

From Sect. 12.1.4 it could be concluded that the measurement of the remanent flux distribution of a superconducting sample is a better test of sample quality than the measurement of the spatial distribution of the levitation force. In fact the scanning Hall probe experiment appears to be a good method for nondestructive testing of superconductors. The measurement of remanent flux distribution in superconductors by a Hall probe has been reported in the literature [152, 154, 221, 274]. At the present time several groups have set up Hall probe experiments to examine the homogeneity of superconducting samples [120, 136, 211, 262]. However, in most cases only magnetic-field distributions are measured, which give a qualitative measure of the homogeneity of the superconductor, while in our experiments a method is proposed to evaluate the critical current density from the remanent flux distribution, yielding a more quantiative determination of the quality of the samples. In the following our own experiments will be described in more detail.

13.1 Experimental Details

The computer-controlled three-dimensional positioning system which was described in Sect. 12.1.1 was used for the scanning Hall probe experiments. As a modification a Hall sensor was used instead of the force sensor.

13.1.1 The Hall Sensor

The Hall sensor has to fulfill several conditions. It has to operate stably and with reasonable output signals at 77 K. Its area should be small enough to obtain a good spatial resolution, and the thickness of the sensor housing should be small in order to achieve a small distance between the Hall sensor and the sample. The latter condition is also necessary for good resolution.

Owing to the small distance between the Hall probe and the sample, which is kept in liquid nitrogen, the sensor becomes cold. The temperature, however, would be very unstable if the sensor was kept above the liquid bath. This is not tolerable, as the Hall voltage is strongly temperature-dependent. Therefore the liquid level was kept high enough for the Hall probe to be immersed in liquid nitrogen, too. Thus stable thermal conditions were guaranteed.

A commercial Hall sensor appeared to be a good choice for our purpose. The physical parameters of the sensor KSY10, manufactured by Siemens, are summarized in Table 13.1.

Table 13.1. Technical data of the Hall probe KSY10

Quantity	Technical data
Hall-active material	Monocrystalline GaAs
Semiconductor size	$200 \times 200\,\mu m^2$
Maximum Hall current I_H	5 mA
Hall constant at 297 K K_{RT}	1.102 ± 0.005 T/V
Hall constant at 77 K K_{LN}	0.81 ± 0.01 T/V
Linearity of the Hall voltage for $B < 0.5$ T	$\leq \pm 0.2\%$

13.1.2 Resolution of the Hall Probe

The trapped fields are well within the region of fields specified for Hall sensors. Thus the electrical output signal (Hall voltage) can easily be measured with standard equipment and with sufficiently low error. The spatial resolution determines the features of the trapped field and thus the size of inhomogeneities of the superconductor that can be resolved by the experiment.

The spatial resolution of the sensor is determined by its active area but even more by the distance between the superconductor and sensor during measurement. In fact the whole area penetrated by the supercurrent acts on the Hall probe. However, areas which are farther away do not contribute as much to the Hall signal as those directly underneath the sensor. One can thus define an effective angle and consider the contribution of areas outside this aperture as not relevant. This is outlined schematically in Fig. 13.1.

From this figure it is clear that a size reduction of the active area of the Hall probe is much less efficient than a reduction of the spacing between the superconductor and sensor. In an actual experiment this may cause a problem. The Hall sensor is mounted in a plastic housing, which owing to its thickness increases the spacing. Thus the housing was ground to reduce its thickness. Furthermore the sensor has to be brought into contact with the sample. Especially in the case of a thin film, this may cause damage of the sample surface. A solution to this problem was found by sealing the film in a transparent foil. The foil was a conventional household material normally used for deep-freezing purposes. Its thickness of 35 μm does not significantly reduce the spacing. Another possibility is spin-coating of the film with photoresist. This reduces the thickness of the protection layer even more.

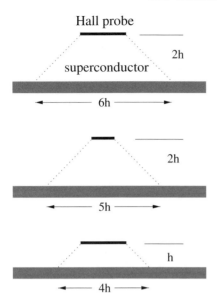

Fig. 13.1. Schematic representation of the spatial resolution as a function of the distance between superconductor and sensor and of the active area of the sensor

13.1.3 Imaging of the Remanent Flux Distribution

The superconducting sample is cooled down to 77 K in the external magnetic field of a copper coil. As soon as this temperature is reached the field is turned off. After a certain time the trapped flux has relaxed and an equilibrium flux distribution, according to the Bean model, is achieved. The normal component of this flux distribution is recorded by scanning a micro-Hall probe over the total surface area of the sample. An ideal cylindrical superconductor, for example, exhibits a perfect Bean cone, whose steepness is a measure of the critical current density. A real sample may show inhomogeneities in the flux distribution caused by sample imperfections such as cracks or grain boundaries where flux can more easily enter and leave the sample. It is possible to calculate the critical current density from this measured flux distribution. The numerical method will be presented in the following.

13.2 Calculation of the Critical Current Density

According to the Bean model the gradient of the magnetic-flux distribution is proportional to the critical current density j_c. In order to calculate the critical current density from the measured flux data the Biot–Savart equation has to be solved. However, the geometrical boundary conditions are not simple; thus a formalism proposed by Xing et al. [274] was applied. A short outline of the calculation will be presented here. A more detailed discussion can be found in [147].

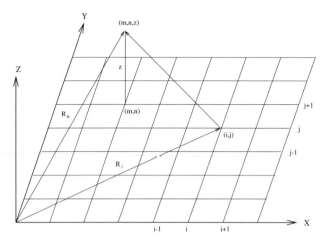

Fig. 13.2. Geometry of the Hall scan

The critical current density is determined from the remanent magnetization:

$$\boldsymbol{j} = \nabla \times \boldsymbol{M}. \tag{13.1}$$

This equation can be reduced to two dimensions. One reason is that with an external field normal to the surface, the screening currents flow only within this plane. The other reason is that the currents should not vary over the thickness of the sample. In particular, in the case of thin films or BSCCO tapes the sample can be considered as thin enough to be treated two-dimensionally. Thus (13.1) can be reduced to two dimensions:

$$\boldsymbol{j} = (j_\mathrm{x}, j_\mathrm{y}) = \left(\frac{\partial M}{\partial y} - \frac{\partial M}{\partial x} \right). \tag{13.2}$$

The geometrical situation for the Hall scan is depicted in Fig. 13.2. The x, y position of the sensor is represented by the integer indices m, n, i, j. The distance between the Hall probe and superconductor is kept constant.

The magnetic moment of a grid point (i, j) induces a remanent magnetic field at the field point (m, n) at a distance z:

$$B_{0,z}(m, n, i, j, z) = \frac{\mu_0}{4\pi} M(i, j) \int_{S_{i,j}} \frac{3z^2 - r^2}{r^5} \mathrm{d}x'\mathrm{d}y'$$
$$= M(i, j)\, G(m, n, i, j, z). \tag{13.3}$$

Here r is the distance between the source point and field point and $S_{i,j}$ is the area of the cell (i, j). The integral G depends on the relative coordinates of the source point and field point. The contributions of all individual cells are now summed. Then the complete induction at the site (m, n, z) is

$$B_z(m,n,z) = \sum_{i=1}^{N_1} \sum_{j=1}^{N_2} M(i,j)\, G(m,n,i,j,z)\,. \tag{13.4}$$

The numbers N_1 and N_2 denote the number of scan steps in the x and y directions and therefore count the total number of data points. Equation (13.4) is a linear system of equations for the magnetization $M(i,j)$ with coefficients $G(m,n,i,j,z)$ and a constant term $B_z(m,n,z)$.

This system of equations can be written as a matrix equation:

$$\boldsymbol{G M} = \boldsymbol{B}_z\,. \tag{13.5}$$

The magnetization $\boldsymbol{M}(i,j)$ can now be calculated either by inversion of the matrix G, which would mean inverting a matrix of $N_1^2 \times N_2^2$ elements, or by applying an iterative method. As the first possibility consumes a lot of computer time, the second was chosen. A comparison of the two methods shows that they yield comparable results. A more detailed discussion of the iteration method is given in [147].

13.3 Epitaxial YBCO Films

Epitaxial thin films of $YBa_2Cu_3O_{7-\delta}$ are currently widely used for microwave and high-frequency applications, for example mobile communication [90]. Films of good quality are produced on large-area substrates up to 8 inches in diameter [246]. In order to fabricate reliable systems it is crucial that the quality of the films is homogeneous over the whole area. A method has to be found which enables nondestructive testing of the overall homogeneity. In our laboratory an inductive method was used [65] but it is restricted in its lateral resolution to the diameter of the coils of 3–4 mm. Also it is very time-consuming; the scan of a 2 in diameter film takes several hours. The scanning Hall probe experiment described above offers a higher lateral resolution (<1 mm) and is much faster.

13.3.1 Experimental Details

The scanning Hall probe measurements were performed as described above. The iteration procedure was used to determine the magnetization from the trapped-flux data and subsequently the critical current distribution was calculated. For the calculation of j_c the exact distance between the superconductor and Hall probe has to be put into the iteration procedure. As already mentioned, however, the thickness x_{eff} of the housing of the sensor is not very well known. The total spacing is $d_{\text{tot}} = d + x_{\text{eff}}$. An error of 0.5 mm in d_{tot} is enough to cause the iteration not to converge. Therefore x_{eff} was determined experimentally by measuring the trapped-flux distribution for different distances d. For each distance x_{eff} was determined for which the iteration procedure converged best. Magnetization data calculated with this value of x_{eff} for different distances have to correspond well. By this procedure consistent magnetizations $M(x, y)$ were found for $x_{\text{eff}} = 0.8 \pm 0.1$ mm.

13.3.2 Results and Discussion

Figure 13.3 shows results from a Hall scan measurement and calculation of a two-inch YBa$_2$Cu$_3$O$_{7-\delta}$ film. The scan steps were $s = 1$ mm wide. A pronounced Bean cone can be seen, which shows that the film is completely penetrated by magnetic flux. The maximum measured flux density is 20 mT. This is only slightly below the maximum value of 25 mT which can be achieved with a 50 mT coil. Figure 13.3a exhibits the measured flux density and Fig. 13.3b the magnetization calculated from the data. Figure 13.3c shows the critical current density distribution which follows from the magnetization of Fig. 13.3b. In the latter a drop of j_c can be seen in the center of the film. This is only in part caused by the numerical method. The numerical "blind spot" should be only a few mm^2; the area of reduction, however, is more than 10 mm^2. There is a physical explanation for this reduced critical current density. The middle of the film is the area with the highest flux density. As for the explanation of the levitation force experiments, the magnetic-field dependence of the critical current density has to be taken into account. With the Kim–Anderson correction (12.11) with $\beta = 1$, the critical current density distribution of Fig. 13.3d is obtained. The variation of j_c is significantly lower. Most of the area of the film has a critical current density of $j_c = 2 - 3$ MA/cm^2. This result corresponds well with the critical current densities of 2.5 MA/cm^2 determined from inductive measurements [34].

The scanning Hall probe experiment is a good means for nondestructive evaluation of high-quality YBa$_2$Cu$_3$O$_{7-\delta}$ films. It could be shown that artificial defects of 1 mm size can be resolved [33]. Magneto-optical measurements can achieve higher lateral resolution [109] but the experimental effort is higher. A very high spatial resolution can be obtained when the Hall probe measurement is combined with scanning probe techniques. The scanning Hall probe microscope [207, 208, 209] yields submicron resolution. The increasing resolution is combined with a more sophisticated experimental setup. For nondestructive evaluation a lower resolution might be sufficient.

13.4 Bi-2223/Ag Tapes

The critical current density of BSCCO tapes and wires is normally measured by a four-point transport experiment. The critical current I_c is defined as the current at a voltage drop of 1 µV over a 1 cm distance between voltage taps. A transport measurement, however, yields only an integral value of the critical current. Local variations can only be detected by using very small

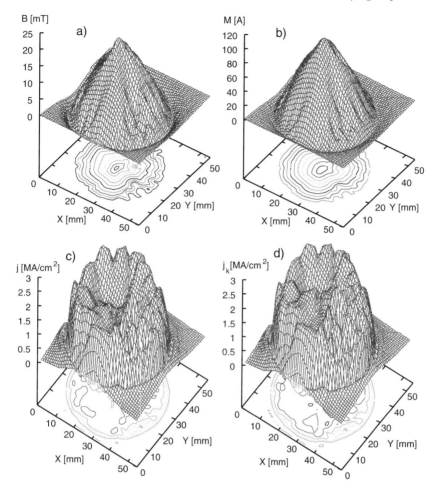

Fig. 13.3. Results of the scanning Hall probe measurements of film MG385: **a)** trapped flux, **b)** magnetization calculated from the trapped flux, **c)** critical current density j_c calculated without and **d)** with Kim–Anderson correction

distances between voltage taps, which is experimentally complicated [222]. Thus the scanning Hall probe experiment turns out to be a good means of nondestructive evaluation of superconducting tapes, too.

13.4.1 Lateral Resolution

The lateral resolution was tested using pure silver tapes. Thus the not-very-well-known influence of the superconductor could be excluded. A transport current was applied and the flux distribution due to that current was recorded at both 77 K and room temperature. Artificial defects of known dimensions were introduced into one silver tape.

13. Remanent Flux Distribution and Critical Current Density

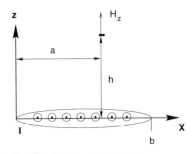

Fig. 13.4. These geometrical parameters of a tape enter the calculation of flux distribution using the Biot–Savart law

The magnetic field of the transport current can be calculated from the Biot–Savart law. For a long, homogeneous tape the problem can be treated two-dimensionally. The z component of the magnetic field thus follows from the Biot–Savart law as

$$H_z(a,h) = \int_0^b dx\, I(x) \frac{a-x}{[(a-x)^2 + h^2]^{3/2}}. \tag{13.6}$$

The parameters of this equation are visualized in Fig. 13.4.

Figure 13.5 shows the flux distribution of a transport current of 8 A through a defect-free silver tape. This flux map corresponds quite well within experimental errors with the distribution of a homogeneous conductor as calculated from (13.6).

The same silver tape was measured again after several holes, cuts and cracks were introduced. The flux map of the tape with defects is shown in Fig. 13.6. It turns out that the method has its greatest resolution at the border of the tape as there the trapped flux has its maximum. Lateral cuts 0.5 mm deep can be resolved. Owing to the lower flux density in the middle of the tape the resolution there is 0.7 mm.

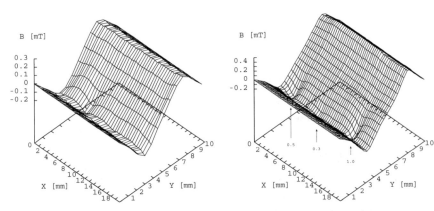

Fig. 13.5. Flux distribution of a defect-free silver tape with a transport current of $I = 8$ A

Fig. 13.6. Flux distribution of a silver tape with defects at a transport current of $I = 8$ A

13.4.2 Results from Bi-2223/Ag Tapes

In the following, results from a typical sample will be presented. It was a Bi-2223/Ag tape with 19 filaments prepared using the PIT technique as described in Chap. 7. Its thickness was 0.25 mm and its width 3.2 mm. For a first impression the critical current was measured using the four-probe technique with 1 cm distance between taps over a total length of 8 cm. The result is shown in Fig. 13.7. Most of the tape had a critical current of more than 30 A. In one region the critical current dropped to 20 A. This small area limits the critical current density over the total length. This area was therefore selected for scanning Hall probe examination.

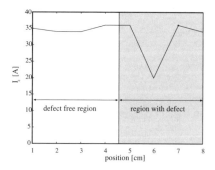

Fig. 13.7. Critical current I_c of a Bi-2223/Ag tape measured at different positions

Two methods of magnetization were used for these experiments. In the first method an external magnetic field was applied during cooling, as described above. The second method was the determination of the flux pattern of a transport current. The region with defects was compared with a defect-free region from the same tape. In the top part of Fig. 13.8 the flux distribution (left) and j_c distribution (right) of the defect-free region after magnetizing in the field of a coil are shown. The bottom part of the figure exhibits the results for the part of the tape with the defect. The flux distribution shows the typical Bean form. The defect can be clearly localized from the reduced magnetic flux and critical current density, reduced from $20\,000\,\text{A/cm}^2$ to $10\,000\,\text{A/cm}^2$, within a range smaller than 3 mm.

Figure 13.9 represents the flux distribution for a transport current of 20 A, and the critical current density distribution calculated from the flux map (top). The current of 20 A lies slightly below the critical current determined by the transport measurement. The bottom part shows the remanent field measured after turning off the transport current, and its critical current distribution. The defect cannot be seen as clearly as in the trapped-flux measurement. However, in the direct neighborhood of the defect an enhancement of the transport current can be observed. This is an indication that the current is flowing around the defect region. The defect thus reduces the effective area for the current path.

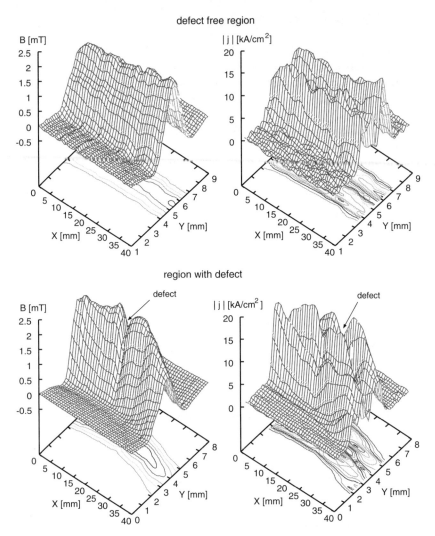

Fig. 13.8. Remanent field and critical current distribution calculated from the flux map of a 19-filament Bi-2223/Ag tape without defects (*top*) and with a defect (*bottom*)

The remanent field after turning off the transport current shows the defect much better. In this case the maximum flux density is at the border of the tape. The remaining supercurrents flow along this border; thus this experimental mode is sensitive to defects at the edge of the tape.

The three experimental modes presented, namely the trapped flux, self-field and remanent field of a transport current, have different current patterns. Thus they are sensitive to different regions of the tape. The trapped-flux dis-

13.4 Bi-2223/Ag Tapes

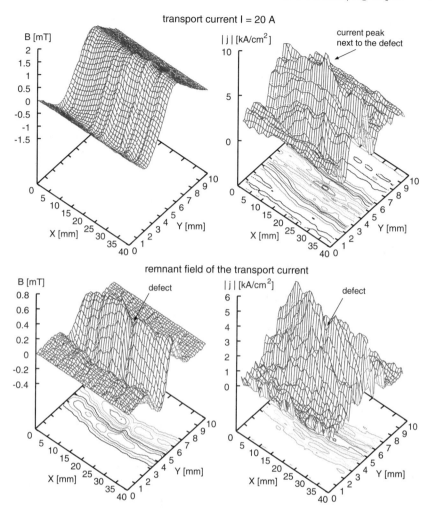

Fig. 13.9. Self-field for a transport current of $I = 20$ A, critical current density distribution calculated from the flux map (*top*) and remanent field of the transport current with the corresponding current distribution (*bottom*) of a 19-filament Bi-2223 tape with a defect

tribution provides a map of the local critical current density but it does not tell us very much about the contacts of the grains. The self-field distribution of a transport current has the advantage that a real current is flowing from one end of the tape to the other. Thus defects which hinder the current path, such as cracks, can be detected. The last mode, in which the remanent-field distribution after a transport current has been applied and turned off is measured, is especially sensitive to the borders of the tape. Thus all three methods are suitable to detect defects in a superconducting tape. Their information is complementary.

13.4.3 Quality Control of Long Tapes

For technical applications long lengths of tapes have to be fabricated. Scanning Hall probe measurements offer a simple means for continuous quality control over the total length. For this purpose it is sufficient to measure in only one direction along the tape length. The tape needs to be magnetized by an external magnetic field before scanning the Hall probe along its length exactly in the middle of the tape. As can be seen from Fig. 13.9, in this mode the remanent flux is highest in the middle and each defect leads to a reduction of the maximum measured flux value. This is visualized in Fig. 13.10, where a cross section of a magnetized tape is represented schematically. A local reduction of the critical current density implies a height reduction of the Bean cone at its summit. Thus a one-dimensional measurement of the remanent flux along the length of a superconducting tape in its middle can reliably detect any bad section of reduced critical current density. An apparatus designed for the purpose of continuous quality control of BSCCO tapes has been demonstrated by Schiller et al. [241].

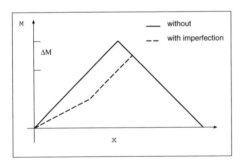

Fig. 13.10. Remanent magnetization of the cross section of a superconducting tape with and without a defect (schematic)

Part V

Concluding Remarks

14. Concluding Remarks

14.1 Summary

The discovery of the oxide superconductors caused a flood of publications. It is therefore difficult for a researcher working in this field to obtain an overview of the latest results and filter the important contributions from the less important ones. For a newcomer, e.g. a student starting his/her master's or PhD thesis work, it is nearly impossible to collect the important information from this vast amount of papers. This situation motivated the first part of the work presented here. The goal was to pick the essential information on technical high-T_c superconductors for magnet and energy technology from several important original articles and books and put it together in a compact form so that students starting in this field would have the opportunity to obtain a quick overview of the basic physics and materials science and a reasonable list of literature. In the second and third part the author's own original results were added. The research was performed at the University of Wuppertal and at the Applied Superconductivity Center of the University of Wisconsin at Madison during a one-year stay and ongoing collaborations.

The experimental results were presented in two parts, one part focusing on phase formation and microstructure, the other on flux-pinning properties measured by electromagnetic methods. In Chap. 7 some general aspects of our preparation processes of Bi-2212 and Bi-2223 conductors were described. In the following Chap. 8 results of overpressure processing of Bi-2212 wires and tapes were presented. Processing in 1 atm oxygen under a total pressure of 5 atm leads to significant reduction of voids in Bi-2212 wires. Tape conductors from the same material processed under the same conditions exhibit a thickness dependence of the critical current density, which is associated with reduced texture for thicker tapes and enhanced second-phase production in very thin tapes (160 µm). The high amount of second phases is attributed to locally inhomogeneous reaction due to the restricted geometry of the tapes.

Second phases also play an important role in the preparation process of Bi-2223 tapes, as described in Chap. 9. It was found that reducing the sintering temperature in the final sintering step leads to significant enhancement of j_c. The reduction of the small amounts of second phases such as Bi-2201 and 3221 could be demonstrated by XRD analysis. Electron microscopy revealed that these phases preferentially reside between grains, reducing the connectivity, as could also be shown by ac susceptibility measurements.

The results on YBCO-coated conductors on textured substrates described in Chap. 10 are very preliminary. It was shown that a good cube texture could be achieved in a Ni–Cu composite material. Buffer layers of CeO_2, YSZ and MgO were sputtered onto textured metal tapes. However, these buffer layers still need to be optimized. The first superconducting YBCO films were deposited onto these substrates with $T_c = 82$ K. This work formed part of the first German activity on coated conductors on textured metallic substrates [6].

Electromagnetic properties were discussed in three chapters. Chapter 11 focused on the scaling behavior of current–voltage characteristics of Bi-2223 tapes in magnetic fields up to 8 T and temperatures down to 4.2 K. The temperature-dependent anisotropy of the critical current density was correlated with the scaling behavior, leading to the conclusion that thermally activated flux creep rather than vortex glass melting is the origin of the critical-current degradation in magnetic fields in the BSCCO conductors. From a technical point of view these results indicate that reasonable applications in magnet design are only possible below 40 K for Bi-2223 and below 30 K for Bi-2212 as the degradation mechanism has to be considered as intrinsic.

Chapter 12 was dedicated to the phenomenon of superconducting magnetic levitation. The main purpose of the experiments presented was to understand the basic physics of the levitation force and magnetic stiffness. Various types of HTSC were examined and the results were discussed in the framework of the Bean critical-state model or the Kim–Anderson model. The phenomenological interpretation of magnetic-stiffness results leads to a qualitative estimation of the Labusch parameter and thus the pinning strength of the samples examined.

In the last chapter, Chap. 13, the measurement of the spatial distribution of trapped magnetic flux was presented. This turned out to be a good method for nondestructive evaluation of various types of HTSC, such as bulk samples, thin films of YBCO and Bi-2223 tapes. A method was presented to calculate the critical-current distribution from the measured flux map. The method yields a spatial resolution of about 1 mm^2.

14.2 Conclusion

The published research work on oxide superconductors can roughly be divided into four types: theoretical work on the understanding of the mechanism of high-T_c superconductivity, fundamental experiments on mechanisms and properties of HTSCs, materials science work and applications. Owing to prospective economical advantages, publishing policies are often restrictive, especially in the fields of materials science and applications. Either experimental details are published incompletely or results are treated as proprietary

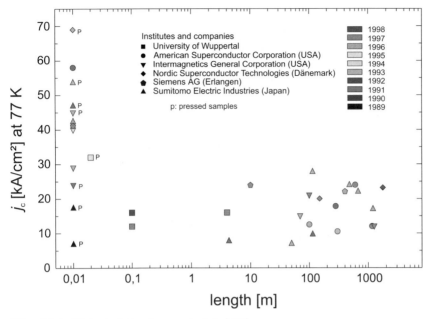

Fig. 14.1. Historical development of j_c [101]

and thus are not made available to the public. This makes it unavoidable that work is done twice. Nevertheless, a comparison with published work is necessary in order to evaluate one's own results. Thus this work has been put into the frame of international results.

From the historical point of view, progress has been slow in the last few years, revealing the problems of material development in these complex compounds. The most important quantity for estimating progress is the critical current density j_c. Thus the Malozemoff plot was introduced [183], in which j_c of Bi-2223/Ag tapes is plotted as a function of time. The critical current density turns out to be linear with time but the plot does not tell the whole story. In Fig. 14.1 the length of the tapes is added as a third parameter.

As can be seen, the highest value of $69\,\mathrm{kA/cm^2}$ was already reached in 1993 in pressed tapes [165]. The most remarkable results in the meantime have been the production of long lengths of tape conductor and the fact that the maximum value is now obtainable by rolling. However, tapes longer than 10 cm still have a j_c about three times lower than that of short samples. This shows the major problems in the field of conductor fabrication. On one hand a reasonable j_c value for commercial applications is considered to be of the order of $10^5\,\mathrm{kA/cm^2}$. On the other hand, the upscaling of critical current densities from short samples to long lengths of conductors necessary to build magnets or cables is very difficult and is connected with a loss in current-carrying capability by a factor of three. Further difficulties arise from the

14. Concluding Remarks

high anisotropy of the critical current and from its degradation in magnetic fields, especially at elevated temperatures. The consequence is that Bi-2223 tapes are applicable in magnet technology only below 40 K and Bi-2212 only below 30 K. At higher temperatures j_c degrades at magnetic fields significantly below 1 T. A way out of this dilemma seemed to be the so-called second-generation coated conductors made of YBCO on various buffer layers (IBAD, RABiTS^{TM}). But upscaling is an even more severe problem in this case. Large-area thick films of YBCO on IBAD buffer layers on polycrystalline metallic substrates have been fabricated [212] but the process is slow and expensive and not necessarily suitable for fabrication of conductors of several km length. The use of rolling-assisted recrystallized metal substrates with appropriate buffer layers could be a solution and a lot of hope is put in this particular type of conductor. But up to now only 10 cm of length have been produced, with significantly degraded critical current density [75]. The reason is that the quality of grain boundaries turns out to be the major problem in these materials as well as in the BSCCO conductors.

As can be seen from Table 14.1 the j_c values obtained in the work presented here are comparable within a factor of two to the results achieved by the leading groups. The engineering current density is about the same owing to the high filling factors which could be reached by the combined deformation techniques described in Sect. 7.1.

Table 14.1. Overview of internationally achieved critical and engineering current densities j_c and j_e (kA/cm^2) of Bi-2223 multifilamentary conductors at 77 K [87, 88, 166, 170, 234, 243, 261]

Institute/company	Short samples		Up to 10 m		More than 10 m	
	j_c	j_e	j_c	j_e	j_c	j_e
University of Wuppertal	32.0	11.4	12.3	3.4	–	–
Sumitomo Electric	42.5	12.1	–	–	27.8 (114 m)	7.9
American Superconductor Corporation	70	15	–	–	24 (600 m)	8
Siemens AG	41.3	10.3	24	6	22 (400 m)	5.5

The results presented in this book were focused mainly on phase formation and microstructure on a micrometer scale, and on electromagnetic properties with a spatial resolution no smaller than 1 mm^2. However, owing to the small coherence length of the HTSCs of only 0.3 to 3 nm, it is clear that impurities and defects on this length scale are most relevant for transport properties. That means that grain boundaries and intragrain properties of this size determine the material properties. Small amounts of second phases visible only in a transmission electron microscope (TEM) can effectively hinder current transport over a grain boundary. Furthermore, the mechanism of current transport over grain boundaries and the type of grain boundary participating in supercurrent transport are still subjects of lively discussion.

Submicron structural features can be found in a TEM but their influence on current transport is not clear from such an experiment. Direct correlations between microstructure and current flow are necessary. Such examination may be done by scanning probe microscopy [175] in combination with microstructural investigations.

The grain boundary problem is not easy to deal with. It may take several years to reach a breakthrough in critical current density or it might even turn out to be impossible to use polycrystalline materials. In contrast to the BSCCO conductors, epitaxial YBaCuO films have already reached high material quality, but these are quasi-single-crystal materials and thus only low-angle grain boundaries are present. The second-generation YBaCuO-coated conductors offer a possible way out of the grain boundary problem. Therefore future research has to include the development of this type of material, too.

This work was funded by the German Ministry of Education, Science, Research and Technology (BMBF) and the Ministry of Science of the state of North Rhine–Westphalia.

References

1. F. Abbatista, et al.: Mater. Chem. Phys. **21**, 521 (1989)
2. A. A. Abrikosov: Sov. Phys. JETP **5**, 1174 (1957)
3. A. A. Abrikosov: *Fundamentals of the Theory of Metals*. North-Holland, Amsterdam, 1988
4. B. L. Adams, S. I. Wright, K. Kunze: Orientation imaging: the emergence of a new microscopy. Metallurg. Trans. A **24**, 219 (1993)
5. P. B. Allen, Z. Fisk, A. Migliori: Normal state transport and elastic properties of high T_c materials and related compounds. In *Physical Properties of High Temperature Superconductors I* (ed. D. M. Ginzberg), pp. 214–264, World Scientific, Singapure, 1989
6. B. Aminov, M. Hortig, S. Kreiskott, B. Lehndorff, H. Piel, J. Pouryamout, N. Pupeter, R. Wagner, D. Wehler: Zwischenbericht 1999 zum Verbundvorhaben YBCO-beschichtete Bandleiter für die Magnet- und Energietechnik. Unpublished (1999)
7. J. W. Anderson, S. E. Dorris, J. A. Parrell, D. C. Larbalestier: The effect of lead content on the critical current density, irreversibility field, and microstructure of Ag-clad $Bi_{1.8}Pb_xSr_2Ca_2Cu_3O_y$ tapes. J. Mater. Res. **14**, 340 (1999)
8. J. W. Anderson, J. A. Parrell, M. Polak, D. C. Larbalestier: Determination of irreversibility field variation on mono- and multifilamentary $(Bi,Pb)_2Sr_2Ca_2Cu_3O_y$ tapes by transport current methods. Appl. Phys. Lett. **71**, 3892 (1997)
9. P. W. Anderson: Phys. Rev. Lett. **9**, 309 (1962)
10. P. W. Anderson, Y. B. Kim: Hard superconductors: theory of the motion of Abrikosov flux lines. Rev. Mod. Phys. **36**, 39 (1964)
11. S. E. Babcock, J. L. Vargas: The nature of grain boundaries in the high-T_c superconductors. Annu. Rev. Mater. Sci. **25**, 193 (1995)
12. J. Bardeen, L. Cooper, J. Schrieffer: Theory of superconductivity. Phys. Rev. **108**, 1175 (1957)
13. J. Bardeen, M. J. Stephen: Phys. Rev. A **140**, 1197 (1965)
14. C. Bean: Magnetization of high field superconductors. Rev. Mod. Phys. **36**, 31 (1964)
15. G. Bednorz, K. A. Müller: Possible high temperature superconductivity in LaSrCuO. Z. f. Physik B **64**, 189 (1986)
16. V. Beilin, A. Goldgirsh, M. Schieber, H. Harel: The blistering phenomenon in BiSCCO-2223 high-T_c superconducting tapes. Supercond. Sci. Technol. **9**, 549 (1996)
17. M. D. Bentzon, P. Vase: Critical current measurement on long BSCCO tapes using a contact-free method. IEEE Trans. Appl. Supercond. **9**, 1594 (1999)

18. G. Blatter, M. V. Feigel'man, V. B. Geshkenbein, A. I. Larkin, V. M. Vinokur: Vortices in high-temperature superconductors. Rev. Mod. Phys. **66**, 1125 (1994)
19. H. J. Bornemann, M. Sander: Conceptual system design of a 5 MWh/100 MW superconducting flywheel energy storage plant for power utility applications. IEEE Trans. Appl. Supercond. **7**, 398 (1997)
20. E. H. Brandt: Rigid levitation and suspension of high-temperature superconductors by magnets. Am. J. Phys. **58**, 43 (1990)
21. E. H. Brandt: Flux line lattice in high-T_c superconductors: anisotropy, elasticity, fluctuation, thermal pinning, AC penetration and susceptibility. Physica C **195**, 1 (1992)
22. E. H. Brandt: The flux-line lattice in superconductors. Rep. Prog. Phys. **58**, 1465 (1995)
23. S. Bröer: *Die Herstellung grossflächiger, einkristalliner $YBa_2Cu_3O_{7-\delta}$-Filme durch planares DC-Hochdrucksputtern.*. Diplomarbeit, Universität Dortmund (1995)
24. W. Buckel: *Supraleitung: Grundlagen und Anwendungen*. Physik-Verlag, Wienheim, 1977
25. R. C. Budhani, J. O. Willis, M. Suenaga, M. P. Maley, J. Y. Coulter, H. Safa, J. L. Ullmann, P. Haldar: Studies of flux pinning by proton induced fission tracks in multifilamentary tapes of $Bi_2Sr_2Ca_2Cu_3O_{10}$/Ag superconductors. J. Appl. Phys. **82**, 3014 (1997)
26. L. N. Bulaevskii, L. L. Daemen, M. P. Maley, J. Y. Coulter: Limits to the critical currrent in high-T_c superconducting tapes. Phys. Rev. B **48**, 13798 (1993)
27. X. Y. Cai, A. Polyanskii, Q. Li, G. N. Riley, D. C. Larbalestier: Current limiting mechanism of individual filament extracted from superconducting tape. Nature **392**, 906 (1998)
28. Z. X. Cai, Y. Zhou, D. O. Welch: Layer rigidity model and the mechanism for ion-diffusion-controlled kinetics in the bismuth cuprate 2212-to-2223 transformation. Phys. Rev. B **52**, 13035 (1995)
29. A. M. Campbell: The response of pinned flux vortices to low-frequency fields. J. Phys. C **2**, 1492 (1969)
30. A. M. Campbell: The interaction distance between flux lines and pinning centres. J. Phys. C **4**, 3186 (1971)
31. A. M. Campbell, J. E. Evvetts: Critical currents in superconductors. Adv. Phys. **72**, 199 (1972)
32. A. D. Caplin, L. F. Cohen, M. N. Cuthbert, M. Dhalle, D. Lacey, G. K. Perkins, I. V. Thomas: Critical current in conductors: exploring the limiting mechanism. IEEE Trans. Appl. Supercond. **5**, 1864 (1995)
33. A. Cassinese, M. Getta, T. Kaiser, A.-G. Kürschner, B. Lehndorff, G. Müller, H. Piel, B. Skriba: Scanning Hall probe measurements on single- and double-sided sputtered YBCO films for microwave applications. IEEE Trans. Appl. Supercond. **9**, 1960 (1999)
34. A. Cassinese, M. Getta, H.-G. Kürschner, B. Lehndorff, G. Müller: Measurement of the Uniformity of HTSC Films by a Scanning Hall Probe. Unpublished (1997)
35. G. W. Castellan: *Physical Chemistry*. Benjamin Cummings, Menlo Park, 1983
36. R. J. Cava, B. Batlogg, R. B. van Dover, D. W. Murphy, S. Sunshine, T. Siegrist, J. P. Remeika, E. A. Rietman, S. Zahurak, G. P. Espinosa: Bulk superconductivity at 93 K in single phase oxygen-deficient perovskite $Ba_2YCu_3O_{9-\delta}$. Phys. Rev. Lett. **58**, 1676 (1987)

37. M. D. Cima, M. C. Flemings, A. M. Figueredo, M. Nakade, H. Ishii, H. D. Brody, J. S. Haggerty: J. Appl. Phys. **71**, 1868 (1992)
38. H. Claus, M. Braun, A. Erb, K. Röhberg, H. Wühl, G. Bräuchle, P. Schweib, G. Müller-Vogt, H. von Löhneisen: The 90 K-plateau of oxygen deficient $YBa_2Cu_3O_{7-\delta}$ single crystals. Physica C **198**, 42 (1992)
39. J. R. Clem: Anisotropic superconductors: fundamentals of vortices in layered superconductors. In *Proc. NATO Adv. Study Inst. "Vortices in Superconductors", Cargese, Corsica* (ed. N. Bontemps), Kluwer, Dordrecht, 1993
40. M. N. Cuthbert, M. Dhalle, J. Thomas, A. D. Caplin, S. X. Dou, Y. C. Guo, H. K. Liu, R. Flükiger, G. Grasso, W. Goldacker, J. Kessler: Transport and magnetisation measurement of Bi-2223/Ag tapes and the role of granularity on critical current limitation. IEEE Trans. Appl. Supercond. **5**, 1391 (1995)
41. M. Däumling, R. Maad, A. Jeremie, R. Flükiger: Phase coexistence and critical temperature of the $Bi,Pb_2Sr_2Ca_2Cu_3O_x$ phase under partial pressure of oxygen between 10^{-3} and 0.12 bar with and without addition of silver. J. Mater. Res. **12**, 1445 (1997)
42. W. I. F. David, et al.: Structure and crystal chemistry of the high-T_c superconductor $YBa_2Cu_3O_{7-x}$. Nature **327**, 310 (1987)
43. S. E. Dorris, B. C. Prorok, M. T. Lanagan, N. B. Browning, M. R. Hagen, J. A. Parrell, Y. Feng, A. Umezawa, D. C. Larbalestier: Physica C **223**, 163 (1994)
44. S. X. Dou, H. K. Liu, M. H. Apperly, K. H. Song, C. C. Sorell: Critical current density in superconducting Bi–Pb–Sr–Ca–Cu–O wires and coils. Supercond. Sci. Technol. **3**, 138 (1990)
45. H. S. Edelman, D. C. Larbalestier: Resistive transition and the origin of the n value in superconductors with a Gaussian critical-current distribution. J. Appl. Phys. **74**, 3312 (1993)
46. O. Eibl: The high-T_c compound $Bi,Pb_2Sr_2Ca_2Cu_3O_{10+\delta}$: features of the structure and microstructure relevant for devices in magnet and energy technology. Supercond. Sci. Technol. **8**, 833 (1995)
47. J. W. Ekin, D. K. Finnemore, Q. Li, J. Tenbrink, W. Carter: Appl. Phys. Lett. **61**, 858 (1992)
48. U. Essmann, H. Träuble: J. Sci. Instr. **43**, 344 (1966)
49. U. Essmann, H. Träuble: Phys. Rev. A **24**, 526 (1967)
50. T. Fahr, W. Pitschke, H.-P. Trinks, K. Fischer: Investigation of the formation of Bi-2223 in Bi-2223/Ag tapes by in-situ X-ray diffraction. Inst. Phys. Conf. Ser. **167(I)**, 587 (2000)
51. B. Fischer: *Herstellung und Charakterisierung silberumhüllter $Bi,Pb_2Sr_2Ca_2Cu_3O_{10}$-Bandleiter und $Bi_2Sr_2CaCu_2O_8$-Drähte*. Dissertation, WUB-DIS 97-9, Bergische Universität GH-Wuppertal (1997)
52. B. Fischer, T. Arndt, J. Gierl, H. Krauth, M. Munz, A. Szulczyk, M. Leghissa, H.-W. Neumüller: Bi-2223 tape processing. Inst. Phys. Conf. Ser. **167(I)**, 463 (2000)
53. K. H. Fischer, T. Nattermann: Collective flux creep in high T_c-superconductors. Phys. Rev. B **43**, 10372 (1991)
54. D. S. Fisher, M. P. A. Fisher, D. A. Huse: Thermal fluctuations, quenched disorder, phase transitions, and transport in type II superconductors. Phys. Rev. B **43**, 130 (1991)
55. R. Flükiger, G. Grasso, J. C. Grivel, F. Marti, M. Dhallé, Y. Huang: Phase formation and critical current density in (Bi,Pb)-2223 tapes. Supercond. Sci. Technol. **10**, A68 (1997)
56. R. Flükiger, A. Jeremie, B. Hensel, E. Seibt, J. Q. Xu, Y. Yamada: presented at the ICMC, Huntsville, USA (1991)

57. S. R. Foltyn, P. N. Arendt, P. C. Dowden, R. F. DePaula, J. R. Groves, J. Y. Coulter, Q. Jia, M. P. Maley, d. E. Petersen: High-T_c coated conductors–performance of meter-long tapes. IEEE Trans. Appl. Supercond. **9**, 1519 (1999)
58. H. C. Freyhardt, J. Hoffmann, J. Weismann, J. Dzich, K. Heinemann, A. Isaev, F. Garcia-Moreno, S. Sievers, A. Usoskin: YBaCuO thick film on planar and curved technical substrates. IEEE Trans. Appl. Supercond. **7**, 1426 (1997)
59. G. Fuchs, P. Stoye, T. Staiger, G. Krabbes, P. Schätzle, W. Gawalek, P. Görnert, A. Gladun: Melt textured YBCO samples for trapped field magnets and levitating bearings. IEEE Trans. Appl. Supercond. **7**, 1949 (1997)
60. H. Fujii, H. Kumakura, H. Kitaguchi, K. Togano, W. Zhang, Y. Feng, E. E. Hellstrom: The effect of oxygen partial pressure during heat treatment on the microstructure of dip-coated Bi-2212/Ag and Ag alloy tapes. IEEE Trans. Appl. Supercond. **7**, 1707 (1997)
61. R. Funahashi, I. Matsubara, K. Ueno, H. Ishikawa, K. Mizuno, N. Ohno: Improvement in critical current density under magnetic field in heavily Pb-doped $Bi_2Sr_2CaCu_2O_x$ superconducting tape. Appl. Phys. Lett. **71**, 1715 (1997)
62. W. Gawalek, T. Habisreuther, T. Strasser, M. Wu, D. Litzkendorf, K. Fischer, P. Görnert, A. Gladun, P. Stoye, P. Verges, K. V. Ilushin, L. K. Kovalev: Remanent induction and levitation force of melt textured YBCO. Applied Superconductivity **2**, 465 (1994)
63. W. Gawalek, T. Strasser, T. Habisreuther, D. Litzkendorf, K. Fischer, P. Görnert: Magnetic characterization of YBCO superconductors for levitation application. In *Applied Superconductivity* (ed. H. C. Freyhardt), p. 969, DGM-Informationsgesellschaft, Oberursel, 1993
64. W. Gawalek, et al.: Preparation and magnetic properties of $YBa_2Cu_3O_{7-x}$ single crystals containing Y_2BaCuO_5 and barium titanate inclusions. Cryogenics **33**, 65 (1993)
65. M. Getta: *Induktive Bestimmung der kritischen Stromdichte und kritischen Temperatur an epitaktischen $YBa_2Cu_3O_{7-\delta}$-Filmen*. Diplomarbeit, WU D 94-15, Bergische Universität GH-Wuppertal (1994)
66. E. Giannini, E. Bellingeri, R. Passerini, R. Flükiger: Direct observation of the Bi,Pb(2223) phase formation inside Ag-sheathhed tapes and quantitative secondary phase analysis by means of in-situ high-temperature neutron diffraction. Physica C **315**, 185 (1999)
67. V. L. Ginsburg, L. D. Landau: K teorii sverkhprovodimosti. Zh. Eksp. Teor. Fiz. **20**, 1044 (1950)
68. B. A. Glowacki, C. J. van der Beck, M. Koncykowski: Dynamic flux patterns of multifilamentary $Ag/Bi_2Sr_2Ca_2Cu_3O_{10-\delta}$. Inst. Phys. Conf. Ser. **167(I)**, 779 (2000)
69. W. Goldacker, B. Ullmann, A. Gäbler, R.Heller: Properties of Bi(2223)/Ag Au multifilamentary tapes for current leads. Inst. Phys. Conf. Ser. **158**, 1223 (1997)
70. von Göler, G. Sachs: Walz- und Rekristallisationstextur regulär-flächenzentrierter kubischer Metalle III. Z. f. Physik **56**, 477 (1929)
71. von Göler, G. Sachs: Walz- und Rekristallisationstextur regulär-flächenzentrierter kubischer Metalle IV. Z. f. Physik **56**, 485 (1929)
72. von Göler, G. Sachs: Walz- und Rekristallisationstextur regulär-flächenzentrierter kubischer Metalle V. Z. f. Physik **56**, 495 (1929)
73. L. P. Gorkov: Sov. Phys. JETP **9**, 1364 (1960)
74. R. Goss Levi: Physics Today **Mai**, 17 (1993)

75. A. Goyal, D. Norton, D. Christen, E. Specht, M. Paranthaman, D. Kroeger, J. Budai, Q. He, F. List, R. Feenstra, H. Kerchner, D. Lee, E. Hatfield, P. Martin, J. Mathis, C. Park: Epitaxial superconductors on rolling-assisted biaxially-textured substrates RABiTS: a route towards high critical current density wire. Appl. Supercond. **4**, 403 (1996)
76. A. Goyal, D. P. Norton, J. D. Budai, M. Paranthaman, E. D. Specht, D. M. Kroeger, D. K. Christen, Q. He, B. Saffan, F. A. List, D. F. Lee, P. M. Martin, C. E. Klabunde, E. Hartfield, V. K. Sikka: High critical current density superconducting tapes by epitaxial deposition of $YBa_2Cu_3O_{7-\delta}$ thick films on biaxially textured metals. Appl. Phys. Lett. **69**, 1795 (1996)
77. D. Grindatto, J.-C. Grivel, G. Grasso, H.-U. Nissen, R. Flükiger: TEM study of the Bi,Pb Sr Ca Cu O phase formation in 2223. Physica C **298**, 41 (1998)
78. D. P. Grindatto, B. Hensel, G. Grasso, H.-U. Nissen, R. Flükiger: TEM study of twist boundaries and colony boundaries in $Bi_2Sr_2Ca_2Cu_3O_{10}$ silver sheathed tapes. Physica C **271**, 155 (1996)
79. J.-C. Grivel, R. Flükiger: Visualization of the formation of the $Bi,Pb_2Sr_2Ca_2Cu_3O_{10+\delta}$ phase. Supercond. Sci. Technol. **9**, 555 (1996)
80. B. Günther: *Induktive Messungen an BSCCO-2223 Bandleitern in niederfrequenten magnetischen Wechselfeldern*. Diplomarbeit, WU D 97-45, Bergische Universität GH-Wuppertal (1997)
81. A. Gurevich, E. A. Pashitskii: Current transport through low-angle grain boundaries in high-temperature superconductors. Phys. Rev. B **57**, 13878 (1998)
82. P. Haasen: *Physikalische Metallkunde*. Springer, Berlin, Heidelberg, 1994
83. Z. Han, P. Bodin, W. Wang, M. D. Bentzon, P. Skov-Hansen, J. Goul, P. Vase: Fabrication and characterization of superconducting Bi-2223 tapes with high critical current densities in km length. IEEE Trans. Appl. Supercond. **9**, 2537 (1999)
84. Z. Han, P. Skov-Hansen, T. Freltoft: The mechanical deformation of superconducting BiSrCaCuO/Ag composites. Supercond. Sci. Technol. **10**, 371 (1997)
85. P. Hardenbicker: *Entwicklung von Bi-2223 Bandleitern zum Bau von Strombegrenzern und Stromzuführungen*. Diplomarbeit, WU D 98-16, Bergische Universität GH-Wuppertal (1998)
86. T. Hasegawa, Y. Hikichi, T. Koizumi, A. Imai, K. Kumakura, H. Kitaguchi, K. Togano: Fabrication and properties of $Bi_2Sr_2CaCu_2O_8$ multilayer superconducting tapes and coils. IEEE Trans. Appl. Supercond. **7**, 1703 (1997)
87. K. Hayashi: Development of Ag-sheathed Bi-2223 superconducting wires and their application. In *Advances in Superconductivity IX*, vol. 2, p. 819, Springer, Tokyo, 1997
88. K. Hayashi, S. Hahakura, N. Saga: Development of Ag-sheathed Bi2223 superconducting wires and their application to magnets. IEEE Trans. Appl. Supercond. **7**, 1229 (1997)
89. M. Hein: *Hochfrequenzeigenschaften granularer Hochtemperatursupraleiter*. Dissertation, WUB DIS 92-2, Bergische Universität GH-Wuppertal (1992)
90. M. A. Hein: *High-Temperature-Superconductor Thin Films at Microwave Frequencies*. Springer, Berlin, Heidelberg, 1999
91. N. F. Heinig, R. D. Redwing, X. Y. Cai, I. F. Tsu, S. E. Babcock, J. E. Nordman, D. C. Larbalestier, D. L. Kaiser: The low to high angle electromagnetic transition in $YBa_2Cu_3O_{7-x}$. In *Proceedings of the 21st International Conference on Low Temperature Physics*, Prague, 1996
92. N. F. Heinig, R. D. Redwing, J. E. Nordman, D. C. Larbalestier: Strong to weak coupling transition in low misorientation angle thin film $YBa_2Cu_3O_{7-x}$ bicrystals. Phys. Rev. B **60**, 1409 (1999)

93. N. F. Heinig, R. D. Redwing, I. F. Tsu, A. Gurevich, J. E. Nordman, S. E. Babcock: Evidence for channel conduction in low misorientation angle [001] tilt YBa$_2$Cu$_3$O$_{7-x}$ bicrystal films. Appl. Phys. Lett **69**, 577 (1996)
94. E. E. Hellstrom: Processing Bi-based high-T_c superconducting tapes, wires and thick films for conductor applications. In *High-Temperature Superconducting Materials Science and Engineering* (ed. D. Shi), p. 383, Elsevier Science, Oxford, 1994
95. B. Hensel, G. Grasso, R. Flükiger: Limits to the critical transport current in superconducting Bi$_2$Sr$_2$Ca$_2$Cu$_3$O$_{10}$ silver sheathed tapes – the railway switch model. Phys. Rev. B **51**, 15456 (1995)
96. S. Hensen: *Hochfrequenzeigenschaften optimierter Hochtemperatursupraleiter Filme*. Dissertation, in preparation, Bergische Universität GH-Wuppertal (1999)
97. S. Hensen, G. Müller, C. T. Rieck, K. Scharnberg: In-plane surface impedance of epitaxial YBa$_2$Cu$_3$O$_{7-\delta}$ films: comparison of experimental data taken at 87 GHz with d- and s-wave models of superconductivity. Phys. Rev. B **56**, 6237 (1997)
98. F. Hill: *Präparation, Mikrostrukturanalyse und Hochfrequenzeigenschaften partiell aufgeschmolzener YBa$_2$Cu$_3$O$_{7-\delta}$-Dickschichten*. Diplomarbeit, WU D 93-7, Bergische Universität GH-Wuppertal (1993)
99. T. G. Holesinger, J. F. Bingert, M. Teplitsky, Q. Li, R. Parrella, M. P. Rupich, G. N. Riley Jr.: Spatial variation in composition in high-j_c Bi-2223 tapes. J. Mater. Res. **15**, 285 (2000)
100. T. G. Holesinger, J. F. Bingert, J. O. Willis, V. A. Maroni, A. K. Fisher, K. T. Wu: The effect of variable oxygen pressure during the Bi-2223 tape processing. J. Mater. Res. **12**, 2046 (1997)
101. M. Hortig: *Experimentelle Untersuchungen der kritischen Ströme von (Bi,Pb)$_2$Sr$_2$Ca$_{n-1}$Cu$_n$O$_x$ Hochtemperatursupraleitern*. Dissertation, WUB-DIS 98-10, Bergische Universität GH-Wuppertal (1998)
102. J. Horvat, S. V. Dou, H. K. Liu, R. Bhasale: Critical current through strong links in Ag/Bi–Sr–Ca–Cu–O superconducting tapes. Physica C **271**, 51 (1996)
103. Y. B. Huang, R. Flükiger: Reducing ac losses of Bi-2223 multifilamentary tapes by oxide barriers. Physica C **294**, 71 (1998)
104. Y. B. Huang, G. Grasso, F. Marti, M. Dhalle, G. Witz, S. Clerc, K. Kwasnitza, R. Flükiger: Low ac losses in Bi(2223) tapes with oxide barrier. Inst. Phys. Conf. Ser. **158**, 1385 (1997)
105. J. R. Hull: Using high-temperature superconductors for levitation applications. Journal of Metals (JOM) **51**, 13 (1999)
106. J. R. Hull: Superconducting bearings. Supercond. Sci. Technol. **13**, R1 (2000)
107. IBS Magnete: Catalogue IBS Magnete, Berlin (1993)
108. Y. Iijima, N. Tanabe, D. Kohno, Y. Ueno: In-plane aligned YBa$_2$Cu$_3$O$_{7-\delta}$ thick films deposited on polycrystalline metallic substrates. Appl. Phys. Lett. **60**, 769 (1992)
109. M. V. Indenboom, T. Schuster, M. R. Koblischka, A. Forkl, H. Kronmüller, L. A. Daraschskii, V. K. Vlasko-Vlasov, A. A. Polyanskii, R. L. Prozorov, V. I. Nikitenko: Study of flux distribution in high T_c single crystals and thin films using magneto-optic techniques. Physica C **209**, 259 (1993)
110. A. Jeremie, K. Alami-Yadri, J.-C. Grivel, R. Flükiger: Bi,Pb(2212) and Bi(2223) formation in the Bi–Pb–Sr–Ca–Cu–O system. Supercond. Sci. Technol. **6**, 720 (1993)
111. J. Jiang, T. C. Shields, J. S. Abell, G. Bushnell-Wye: Investigation of phase and texture evolution during thermomechanical processing of Bi(Pb)-2223/Ag tapes by high energy synchrotron X-ray diffraction. Physica C **306**, 91 (1998)

112. K. Jikihara, K. Watazawa, H. Mitsubori, J. Sahuraba, Y. Sugizaki, T. Hasabe, H. Okuba, M. Ichihara, K. Watanabe: A cryocooler cooled 5T superconducting magnet with a horizontal and vertical room temperature bore. IEEE Trans. Appl. Supercond. **7**, 423 (1997)
113. S. Jin, et al.: Appl. Phys. Lett. **52**, 2074 (1988)
114. J. D. Jorgensen, B. W. Veal, A. P. Paulikas, L. J. Novicki, G. W. Crabtree, H. Claus, W. K. Kwok: Structural properties of oxygen deficient $YBa_2Cu_3O_{7-\delta}$. Phys. Rev. B **41**, 1863 (1990)
115. A. W. Kaiser, H. J. B. R. Koch: Processing technique for fabrication of advanced YBCO bulk materials for industrial application. Inst. Phys. Conf. Ser. **158**, 837 (1997)
116. S. S. Kalsi, D. Aized, B. Connor, G. Snitchler, J. Campbell, R. E. Schwall, J. Kellers, T. Stephanblome, A. Tromm, P. Winn: HTS SMES magnet design and test results. IEEE Trans. Appl. Supercond. **7**, 971 (1997)
117. H. Kamerlingh Onnes: The liquefaction of helium. Proc. Roy. Acad. Amsterdam **11**, 168 (1908)
118. H. Kamerlingh Onnes: Further experiments with liquid helium. C: on the change of electric resistance of pure metals at very low temperatures etc. IV: the resistance of pure mercury at helium temperatures. Comm. Phys. Lab. Univ. Leiden **120b** (1911)
119. T. Kanai, N. Inoue: Development of gas pressure melting GPM method for Ag-sheathed Bi-2212 wires. J. Mater. Sci. **30**, 3200 (1995)
120. K. Kawano, A. Oota: A study of self field distribution in Ag-sheathed $Bi.Pb_2Sr_2Ca_2Cu_3O_x$ monofilamentary tape using a scanning Hall probe magnetometry. Physica C **275**, 1 (1997)
121. C.-J. Kim, K. B. Kim, G.-W. Hong: Y_2BaCuO_5 morphology in melt-textured Y–Ba–Cu–O oxides with PtO_2–H_2O/CeO_2 addition. Physica C **232**, 163 (1994)
122. H. Kitaguchi, K. Itoh, T. Takeuchi, H. Kumakura, H. Miao, H. Wada, K. Togano, T. Hasegawa, T. Koizumi: Performance of 10–50 K of Bi-2212/Ag multilayer tape fabricated by using pre-annealing and intermediate rolling process. Physica C **320**, 253 (1999)
123. H. Kitaguchi, H. Kumakura, K. Togano, H. Miao, T. Hasegawa, T. Koizumi: Bi-2212/Ag multilayer tapes with j_c(4.2K,10T) of 500,000 A/cm^2 by using PAIR process. IEEE Trans. Appl. Supercond. **9**, 1794 (1999)
124. H. Kitaguchi, H. Miao, H. Kumakura, K. Togano: Relationship between Bi-2212 layer thickness and j_c enhancment by pre-annealing and intermediate rolling process. Physica C **320**, 71 (1999)
125. C. Kittel: *Einführung in die Festkörperphysik*. Oldenbourg, München, 1976
126. M. Kiuchi, K. Noguchi, T. Matsushita, T. Kato, T. Hikata, K. Sato: Scaling of current–voltage curves in superconducting Bi-2223 silver sheathed tape wires. Physica C **278**, 62 (1997)
127. M. Kiuchi, K. Noguchi, T. H. T. Matsushita, T. Kato, K. Sato: In *Advances in Superconductivity VIII* (eds. H. Hayakawa, Y. Enomoto), p. 615, Springer, Tokyo, 1996
128. N. Klein, et al.: Phys. Rev. Let. **71**, 3355 (1993)
129. R. Kleiner, P. Müller: Intrinsic Josephson effect in layered superconductors. Physica C **293**, 156 (1997)
130. S.-I. Kobayashi, T. Kaneko, T. Kato, J. Fujikami, K.-I. Sato: A novel scaling of magnetic field dependencies of critical currents for Ag-sheated Bi-2223 superconducting tape. Physica C **258**, 336 (1996)
131. J. Kober, et al.: Vibrating-reed experiments on superconducting suspensions. Phys. Rev. Lett. **66**, 2507 (1991)

132. M. R. Koblischka, T. H. Johansen, H. Bratsberg, P. Vase: Flux patterns of monofilamentary $Bi_2Sr_2Ca_2Cu_3O_x$. Supercond. Sci. Technol. **12**, 113 (1999)
133. M. R. Koblischka, T. Schuster, H. Kronmüller: Flux penetration in granular $YBa_2Cu_3O_{7-x}$. Physica C **219**, 205 (1994)
134. R. M. Koch, V. Foglietti, W. J. Gallagher, G. Koren, A. Gupta, M. P. A. Fisher: Experimental evidence for vortex-glass superconductivity in Y–Ba–Cu–O. Phys. Rev. Lett. **63**, 1501 (1989)
135. P. Komarek: *Hochstromanwendungen der Supraleitung*. Teubner Studienbücher, Stuttgart, 1995
136. P. Kovac, V. Cambel, D. Gregusova, P. Elias, I. Husek, R. Kudela, S. Hasenöhrl, M. Durica: Testing the homogeneity of Bi2223/Ag tapes by a Hall probe array. Inst. Phys. Conf. Ser. **158**, 1311 (1997)
137. S. Kreiskott: *Herstellung oxidischer Pufferschichten auf texturierten Substraten*. Diplomarbeit, WU D 99-5, Bergische Universität GH-Wuppertal (1999)
138. J. Krelans, R. Nast, H. Eckelmann, W. Goldacker: Novel, internally stranded ring bundled barrier Bi-2223 tapes for low ac loss application. Inst. Phys. Conf. Ser. p. 467 (2000)
139. V. Z. Kresin, H. Morowitz, S. A. Wolf: *Mechanisms of Conventional and High T_c Superconductivity*. Oxford University Press, Oxford, 1993
140. V. Z. Kresin, S. A. Wolf: Physica C **169**, 476 (1990)
141. Kugler: Private communication (1998)
142. M. Kuhn, B. Schey, W. B. B. Stritzker, J. Eisenmenger, P. Leiderer: Large area magneto-optical investigation of YBCO thin films. Rev. Sci. Instrum. **70**, 1761 (1999)
143. H. Kumakura, H. Kitaguchi, T. Kiyoshi, K. Inoue, K. Togano, M. Okada, F. Fuhushima, K. Tanaka, K. Kato, J. Sato: Performance test of Bi-2212 insert magnets fabricated by Ag sheath method and coating method. IEEE Trans. Appl. Supercond. **7**, 646 (1997)
144. P. Kummeth, C. Struller, H.-W. Neumüller, G. Ries, M. Kraus, M. Leghissa, G. Wirth, J. Wiesner, G. Saemann-Ischenko: The influence of columnar defects on the critical current density in $Bi,Pb_2Sr_2Ca_2Cu_3O_{10+\delta}$/Ag tapes. J. Supercond. **7**, 783 (1994)
145. P. Kummeth, C. Struller, H.-W. Neumüller, G. Saemann-Ischenko: Dimensionality of flux pinning in $Bi,Pb_2Sr_2Ca_2Cu_3O_{10+\delta}$/Ag tapes with columnar defects – crossover from two-dimensional to three-dimensional behaviour. Appl. Phys. Lett. **65**, 1302 (1994)
146. H. Küpfer, et al.: Investigation of inter- and intragrain critical currents in high T_c ceramic superconductors. Cryogenics **28**, 650 (1988)
147. H.-G. Kürschner: *Levitationsmessungen und Hall-Kartographie an Hochtemperatursupraleitern*. Dissertation, WUB-DIS 99-12, Bergische Universität GH-Wuppertal (1999)
148. K. Kwasnitza, S. Clerc: Specific aspects of ac losses in high-T_c superconductors. Adv. Cryogenic Eng. **40**, 53 (1994)
149. K. Kwasnitza, S. Clerc, R. Flükiger, Y. B. Huang: Alternating magnetic field losses in high-T_c superconducting. Physica C **299**, 113 (1998)
150. K. Kwasnitzy, S. Clerc: AC-losses of superconducting high-T_c multifilament Bi-2223/Ag sheathed tape in perpendicular magnetic fields. Physica C **233**, 423 (1994)
151. R. Labusch: Crystal Lattice Defects **1**, 1 (1969)
152. M. Lahtinen, J. Paasi, J. Sarkaniemi, Z. Han, T. Feltoft: Homogeneity study of Bi-2223/Ag monofilamentary tapes using Hall sensor magnetometry. Physica C **244**, 115 (1995)

153. D. C. Larbalestier: Critical current limiting mechanisms in polycrystalline high temperature superconductor, conductor like forms. TCSUH 10th Anniversary Workshop, Houston, Texas (1996)
154. D. C. Larbalestier, X. Y. Cai, Y. Feng, H. Edelman, A. Umezawa, G. N. Riley Jr., W. L. Carter: Position sensitive mesurement of the local critical current density in silver sheathed high temperature superconductor Bi.$Pb_2Sr_2Ca_2Cu_3O_y$ tapes. Physica C **221**, 299 (1994)
155. A. I. Larkin, Y. N. Ovchinnikov: Pinning in type II superconductors. J. Low Temp. Phys. **34**, 409 (1979)
156. W. E. Lawrence, S. Doniach: In *Proc. 11th Int. Conf. Low Temp. Phys. (ed. E. Kanda)*, p. 361, Academic Press of Japan, Kyoto, 1971
157. D. F. Lee, R. G. Presswood Jr., K. Salama, T. L. Francavilla, M. Eisterer, H. W. Weber, S. Zanella: Processing and properties of seeded directional solidified YBCO superconductors for high current application (1994)
158. B. Lehndorff, D. Busch, R. Eujen, B. Fischer, H. Piel, R. Theisejans: Preparation and characterization of silver-sheathed BSCCO-2223 tapes. IEEE Trans. Appl. Supercond. **5**, 1251 (1995)
159. B. Lehndorff, P. Hardenbicker, M. Hortig, H. Piel: Evidence for enhanced grain connectivity due to second phase reduction of Bi-2223/Ag tapes. Physica C **312**, 105 (1999)
160. B. Lehndorff, H.-G. Kürschner, B. Lücke, H. Piel: Levitation force and field mapping of melt processed $YBa_2Cu_3O_{7-x}$ ceramics. Physica C **247**, 280 (1995)
161. B. Lehndorff, H.-G. Kürschner, H. Piel: Mapping of magnetic force and field distribution of melt-textured Y–Ba–Cu–O. IEEE Trans. Appl. Supercond. **5**, 1814 (1995)
162. M. Lehndorff: *Brillouinstreuung an thermischen und mikrowellen-induzierten Fluktuationen in Lithium-dotiertem Kalium-Tantalat*. Series: Konstanzer Dissertationen, Hartung und Gorre, Konstanz, Universität Konstanz (1986)
163. M. Lelovic, P. Krishnaraj, N. G. Eror, A. N. Iyer, U. Balachandran: Transport critical current above 10^5 A/cm^2 at 77 K in the thin layer of Bi-2223 superconductor near the Ag in Ag-sheathed tapes. Supercond. Sci. Technol. **9**, 201 (1996)
164. M. Lenkens: *Optimierung von gesputterten $YBa_2Cu_3O_{7-x}$-Filmen für Hochfrequenzanwedungen*. Dissertation, WUB-DIS 96-5, Bergische Universität GH-Wuppertal (1996)
165. Q. Li, K. Brodersen, H. Hjuler, T. Freltoft: Critical current density enhancement in Ag-sheathed Bi-2223 superconducting tapes. Physica C **217**, 360 (1993)
166. Q. Li, T. G. Holesinger, X. Y. Cai, G. N. Riley, J. Parrell, S. Fleshler, M. W. Rupich, A. E. Polyanskii, A. P. Malozemoff, D. C. Larbalestier: Critical current density enhancement in rolled multifilament Bi-2223 HTS composites. In *1997 International Workshop on Superconductivity (3rd Joint ISTEC/MRS Workshop).*, ISTEC – International Superconductivity Technology Center, Tokyo, (1997)
167. Q. Li, M. Suenaga, T. Kaneko, K. Sato, C. Simmon: Collapse of irreversible field of superconducting $Bi_2Sr_2Ca_2Cu_3O_{10+\delta}$/Ag tapes with columnar defects. Appl. Phys. Lett. **71**, 1561 (1997)
168. Q. Li, M. Suenaga, Q. Li, T. Freltoft: Sample size effect on the determination of the irreversibility line of high-T_c superconductors. Appl. Phys. Lett. **64**, 250 (1994)

169. Q. Li, H. Wiesmann, M. Suenaga, L. Motovidlo, P. Haldar: Observation of vortex-glass liquid transition in the high T_c superconductor $Bi_2Sr_2Ca_3Cu_3O_{10}$. Phys. Rev. B **50**, 6493 (1994)
170. Q. Li, et al.: Progress in superconducting performance of rolled multifilament Bi-2223 HTS composite conductor. IEEE Trans. Appl. Supercond. **7**, 2026 (1997)
171. F. Lindemann: Phys. Z. Leipzig **11**, 69 (1910)
172. D. Litzkendorf, T. Habisreuther, M. Zeisberger, O. Surzhenko, W. Gawalek: Preparation of melt textured YBCO with optimal shape for cryomagnetic application. Inst. Phys. Conf. Ser. **167(I)**, 91 (2000)
173. J. Löhle: *Hochtemperatursupraleitende Bänder und Drähte*. Disseration, Eidgenössische Technische Hochschule Zürich (1993)
174. F. London, H. London: The electromagnetic equation of the superconductor. Proc. Roy. Soc. **A149**, 71 (1935)
175. A. de Lozanne: Scanning probe microscopy of superconductors. Supercond. Sci. Technol. **13**, 1 (1999)
176. B. Lücke: *Untersuchungen zur Levitationskraft und zur remanenten Flussdichte an $YBa_2Cu_3O_{7-x}$-Hochtempteratursupraleitern*. Diplomarbeit, WU D 95-11, Berigsche Universität GH-Wuppertal (1995)
177. B. Lücke, H.-G. Kürschner, B. Lehndorff, M. Lenkens, H. Piel: Levitation force of a stack of epitaxial $YBa_2Cu_3O_{7-\delta}$ films. Physica C **259**, 151 (1996)
178. Y. Luo, A. H. Morish, Q. A. Pankhurst, G. H. Pelletier, G. J. Roy, X. Z. Zhou: New high temperature superconductor Y–Ba–Cu–O. Can. J. Phys. **65**, 438 (1987)
179. H. Maeda, Y. Tanaka, M. Fukumoto, T. Asano: A new high T_c oxide superconductor without rare earth element. Jpn. J. Appl. Phys. **27**, L209 (1988)
180. M. Maeda, et al.: Jpn. J. Appl. Phys. **28**, 1417 (1989)
181. P. Majewski: BiSrCaCuO High-T_c superconductors. Adv. Mater. **6**, 460 (1994)
182. P. Majewski: Phase diagram studies in the system Bi–Pb–Sr–Cu–O–Ag. Supercond. Sci. Technol. **10**, 453 (1997)
183. A. Malozemoff: Talk given at the Applied Superconductivity Conferenc 1996 at Pittsburgh, PA
184. A. P. Malozemoff: In *High Temperature Superconducting Compounds II* (eds. S. H. Whang, et al.), p. 3, TMS, Warrendale PA, 1990
185. L. Martini: Silver-sheathed Bi-2223 tapes: the state of the art. Supercond. Sci. Technol. **11**, 231 (1998)
186. J. E. Mathis, A. Goyal, D. F. Lee, F. A. List, M. Paranthaman, D. K. Christen, E. D. Specht, D. K. Kroeger, P. M. Martin: Biaxially textured $YBa_2Cu_3O_{7-\delta}$ conductors on rolling assisted biaxially textured substrates with critical current densities of $2-3\,MA/cm^2$. Jpn. J. Appl. Phys. **37**, L1379 (1998)
187. T. Matsushita: Pinning force density in strongly pinned superconductors. Physica C **243**, 312 (1995)
188. B. T. Matthias: Phys. Rev. **97**, 74 (1955)
189. B. T. Matthias, T. H. Geballe, S. Geller, E. Corenzwit: Superconductivity of Nb_3Sn. Phys. Rev. **95**, 1435 (1954)
190. R. W. McCallum, K. W. Dennis, L. Margulies, M.-J. Kramer: In *Processing of long length of superconductors* (eds. U. Balachandran, E. W. Collings, A. Goyal), p. 195, TMS, Warrendale, PA, 1994
191. J. Meins, L. Miller: The high speed MAGLEV transportation system TRANSRAPID. IEEE Trans. Magn. **24**, 808 (1988)
192. W. Meissner, R. Ochsenfeld: Ein neuer Effekt bei Eintritt der Supraleitfähigkeit. Naturwissensch. **21**, 787 (1933)

193. H. Miao, H. Kitaguchi, H. Kamakura, K. Togano, T. Hasegawa: Effects of PAIR pre-annealing and intermediate rolling process on superconducting properties of $Bi_2Sr_2CaCu_3O_x$/Ag multilayer tapes. Physica C **301**, 116 (1998)
194. H. Miao, H. Kitaguchi, H. Kimakura, K. Togano, T. Hasegawa, T. Koizumi: $Bi_2Sr_2CaCu_2O_x$/Ag multilayer tapes with $J_c > 500\,000\,A/cm^2$ at 4.2 K and 10 T by using pre-annealing and intermediate rolling process. Physica C **303**, 81 (1998)
195. H. Miao, H. Kitaguchi, H. Kumakura, K. Togano, T. Hasegawa, T. Koizumi: Optimization of melt-processing temperature and period to improve critical current density of Bi-2212/Ag multilayer tapes. Physica C **320**, 77 (1999)
196. C. Michel, M. Herzieu, M. M. Borel, A. Grandin, F. Deslanden, J. Proost, B. Raveau: Z. Phys. B **68**, 421 (1987)
197. A. R. Moodenbaugh, D. A. Fischer, Y. Z. Wang, Y. Fukumoto: Superconductivity, oxygen content, and hole state density in $Bi_2Sr_{1.75}Ca_{1.25}Cu_2O_{8.18+y}$-$0.09 < y \leq 0$ and $Bi_{1.6}Pb_{0.4}Sr_{1.9}Ca_2Cu_3O_z$. Physica C **268**, 107 (1996)
198. F. C. Moon: Magnetic forces in high-T_c superconducting bearings. Appl. Electromagn. Mater. **1**, 29 (1990)
199. C. G. Morgan, B. M. Henry, C. J. Eastell, M. J. Goringe, C. R. M. Grovenor, J. W. Burgoyne, D. Dew-Hughes, M. Priestnall, R. Storey, H. Jones: Continuous melt processing of Bi-2212/Ag dip-coated tapes. IEEE Trans. Appl. Supercond. **7**, 1711 (1997)
200. H. G. Müller: Über die Erholung und Rekristallisation von kaltbearbeitetem Nickel. Z. f. Metallkunde **31**, 161 (1939)
201. J. Müller: *Entwicklung von silberumhüllten Bi-2212 Drähten und Bi-2223 Bandleitern*. Diplomarbeit, Universität Bonn (1996)
202. K.-A. Müller: Talk given on the ocassion of the 75th birthday of Prof. H. E. Bömmel, Konstanz (1987)
203. M. Murakami: Progress in applications of bulk high temperature superconductors. Inst. Phys. Conf. Ser. **167(I)**, 7 (2000)
204. M. Murakami, M. Morita, K. Doi, K. Miyamoto: Jpn. J. Appl. Phys. **28**, 1189 (1989)
205. T. Muroga, J. Sato, H. Kitaguchi, H. Kumakura, K. Togano, M. Okada: Enhancement of critical current density for Bi-2212/Ag conductor through microstructure control. Physica C **309**, 236 (1998)
206. M. Onoda, A. Yamamoto, E. Takayame-Muromachi, S. Takewawa: Assignment of the powder X-ray diffraction pattern of superconductor $Bi_2Sr_2Ca_{3-x}Cu_2O_y$. Jpn. J. Appl. Phys. **27**, L833 (1988)
207. A. Oral, J. C. Barnard, S. J. Bending, I. I. Kaya, S. Ooi, T. Tamegai, M. Henini: Direct observation of melting of the vortex solid in $Bi_2Sr_2CaCu_2O_{8+x}$ single crystals. Phys. Rev. Lett. **80**, 3610 (1998)
208. A. Oral, S. J. Bending, M. Henini: Real time scanning Hall probe microscopy. Appl. Phys. Lett. **69**, 1324 (1996)
209. A. Oral, S. J. Bending, M. Henini: Scanning Hall probe microscopy of superconductors and magnetic materials. J. Vac. Sci. Technol. B **14**, 1202 (1996)
210. S. Orbach-Werbig: *Oberflächenimpedanz epitaktisch aufgewachsener $YBa_2Cu_3O_{7-x}$-Filme bei 87 GHz*. Dissertation, WUB-DIS94-9, Bergische Universität GH-Wuppertal (1994)
211. J. Paasi, T. Kalliohaka, A. Korpela, L. Söderlund, P. F. Herrmann, J. Kvitkovic, M. Majoros: Homogeneity studies of multifilamentary BSCCO tapes by three-axis Hall sensor magnetometry. IEEE Trans. Appl. Supercond. **9**, 1598 (1999)

212. V. M. Pan, A. L. Kosatkin, V. L. Svetchnikov, V. A. Komashko, A. G. Popov, A. Y. Galkin, H. Freyhardt, M. Zeisberger: Critical current density in highly biaxially-oriented YBCO films: can we control $j_c(77\,\text{K})$ and optimize up to more than 10^6 Amp/cm^2? IEEE Trans. Appl. Supercond. **9**, 1535 (1999)
213. C. Park, D. Norton, D. Christen, D. Verebelyi, R. Feenstra, J. Budai, D. Lee, A. Goyal, E. Specht, D. Kroeger, M. Paranthaman: Long length fabrication of YBCO on rolling assisted biaxially textured substrates RABiTS using laser deposition. IEEE Trans. Appl. Supercond. **9**, 2276 (1999)
214. J. A. Parrell, D. C. Larbalestier, G. N. Riley, Q. Li, R. D. Parrella, M. Teplinsky: Enhancement of the 77 K irreversibility field and critical current density of Bi,Pb$_2$Sr$_2$Ca$_2$Cu$_3$O$_y$ tapes by manipulation of the final cooling rate. Appl. Phys. Lett. **69**, 2915 (1996)
215. J. A. Parrell, A. A. Polyanskii, A. E. Pashitzki, D. C. Larbalestier: Direct evidence for residual, preferentially-oriented cracks in rolled and pressed Ag-clad BSCCO-2223 tapes and their effect on the critical current density. Supercond. Sci. Technol. **9**, 393 (1996)
216. A. Pashitski, A. Gurevich, J. Parrell, D. Larbalestier: Magnetic granularity, percolation and preferential current flow in a silver sheathed Bi$_{1.8}$Pb$_{0.4}$Sr$_2$Ca$_2$Cu$_3$O$_{8-x}$ tape. Physica C **246**, 133 (1995)
217. W. Paul, M. Lakner, I. Rymer, P. Untermährer, T. Baumann, M. Chen, L. Widenborn, A. Gudrig: Test of a 1.2 MVA high T_c superconducting current limiter. Inst. Phys. Conf. Ser. **158**, 981 (1997)
218. D. Petrisor, V. Boffa, G. Celentano, L. Ciontea, F. Fabbri, U. Gambardella, S. Ceresara, P. Scardi: Development of biaxially aligned buffer layers on Ni and Ni-based alloy substrates for YBCO tapes fabrication. IEEE Trans. Appl. Supercond. **9**, 2256 (1999)
219. H. Piel, D. Busch, B. Fischer, B. Lehndorff, R. Theisejans: Transport properties of silver sheathed BSCCO-2223 tapes in high magnetic fields. In *Advances in Superconductivity VIII*, vol. 2, p. 757, Springer, Tokyo, 1995
220. M. Polak, M. Majoros, A. Kasztler, H. Kirchmayr: Filament bridging and critical current density variation in the cross-section. IEEE Trans. Appl. Supercond. **9**, 2151 (1999)
221. M. Polak, M. Majoros, J. Kvitkovic, P. Kottmann, P. Kovac, T. Melisek: Magnetic field in the vicinity of BSCCO tapes carrying transport currrent. Cryogenics **34 ICEC Suppl.**, 805 (1994)
222. M. Polak, W. Zhang, J. Parrell, X. Y. Cai, A. Polyanskii, E. E. Hellstrom, D. C. Larbalestier, M. Majoros: Current transfer lengths and the origin of linear components in the voltage–current curves of Ag-sheathed BSCCO components. Supercond. Sci. Technol. **10**, 769 (1997)
223. A. A. Polyanskii, X. Y. Cai, D. M. Feldmann, D. C. Larbalestier: Visualization of magnetic flux in magnetic materials and high temperature superconductors using Faraday effect in ferriagnetic garnet films. In *Ferrimagnetic Nanocrystalline and Thin Film Magnetooptical and Microwave Material* (eds. M. Ausloos, I. Nedkov), NATO ASI Series, Kluweer Academic, Dordrecht, 1999
224. Y. G. Ponomarev, N. B. Brandt, C. S. Khi, S. V. Tchesnokov, E. B. Tsokur, A. V. Yarigin, K. T. Yusupov, B. A. Aminov, M. A. Hein, G. Müller, H. Piel, D. Wehler, V. Z. Kresin, K. Rosner, K. Winzer, T. Wolf: Manifestation of a clear gap structure from point-contact and tunneling spectroscopy of YBa$_2$Cu$_3$O$_{7-x}$ and YbBa$_2$Cu$_3$O$_{7-x}$ single crystals. Phys. Rev. B **52**, 1352 (1995)
225. A. M. Portis: Magnetic Forces on Superconductors. Internal report (1994), Universität Wuppertal

226. H. F. Poulsen, T. Frello, N. H. Anderson, M. D. Betzon, M. von Zimmermann: Structural studies of BSCCO/Ag tapes in high-energy synchrotron X-ray diffraction. Physica C **298**, 265 (1998)
227. H. F. Poulsen, L. Gottschalck-Anderson, T. Frello, S. Prantontep, N. H. Andersen, S. Garbe, I. Madsen, A. Abrahamsen, M. D. Bentzon, M. von Zimmermann: In-situ study of equlibrium phenomena and kinetics in a BSCCO/Ag tape. Physica C **315**, 254 (1999)
228. Qing He, D. K. Christen, J. D. Budai, E. D. Specht, D. F. Lee, A. Goyal, D. P. Norton, M. Paranthaman, F. A. List, D. M. Kroeger: Deposition of biaxially-oriented metal and oxide buffer-layer films on textured Ni tapes: new substrates for high-current high-temperature superconductors. Physica C **275**, 155 (1997)
229. M. Quilitz, W. Goldacker: Oxygen exchange in Bi(2223) tapes with Ag and alloyed AgMg sheaths monitored by a thermogravimetrical relaxation method. Supercond. Sci. Technol. **11**, 577 (1998)
230. V. Randle: *Electron backscatter diffraction*. Oxford, England, 1996
231. J. L. Reeves, M. Polak, W. Zhang, E. E. Hellstrom, S. E. Babcock, D. C. Larbalestier, N. Inoue, M. Okada: Overpressure processing of Ag-sheathed Bi-2212 tapes. IEEE Trans. Appl. Supercond. **7**, 1541 (1997)
232. A. Riise, T. H. Johansen, H. Bratsberg: The vertical magnetic force and stiffness between a cylindrical magnet and a high-T_c superconductor. Physica C **234**, 108 (1994)
233. G. N. Riley, A. P. Malozemoff, Q. Li, S. Fleshler, T. G. Holesinger: The freeway model: new concepts in understanding supercurrent transport in Bi-2223 tapes. Journal of Metals (JOM) **49**, 24 (1997)
234. M. W. Rupich, et al.: Critical current density enhancement in Bi-2223 composites. In *1998 Int. Workshop on Superconductivity*, Okinawa, Japan, 1998
235. K. Salama, A. S. Parikh, P. Putman, L. Woolf: A novel approach in high rate melt-texturing in 123 superconductors. Proc. 10th Anniversary HTS Workshop on Physics, Materials and Applications, Houston, Texas (1996)
236. K. Salama, V. Selvamanickam, L. Gao, K. Sun: Appl. Phys. Lett. **54**, 2352 (1989)
237. K. Salama, V. Selvamanickam, D. F. Lee: Melt-processing and properties of Y–Ba–Cu–O. In *Processing and Properties of High T_c Superconductors* (ed. S. Jin), World Scientific, Singapore, 1992
238. K. Sato, K. Hayashi, K. Ohmatsu, J. Fujikami, N. Saga, T. Shibata, S. Isojima, S. Honjo, H. Ishii, T. Hara, Y. Iwata: HTS large scale application using BSCCO conductor. IEEE Trans. Appl. Supercond. **7**, 345 (1997)
239. K. Sato, T. Hikatan, Y. Iwasa: Critical currents of superconducting BiPbSrCaCuO tapes in the magnetic flux density range 0 – 19.75 T at 4.2, 15 and 20 K. Appl. Phys. Lett. **57**, 1928 (1990)
240. P. Schätzle, A. Berning, W. Bieger, U. Krabbes: Preparation and melt processing of (RE,Y)BaCuO (RE = Nd,Sm) and (Nd,Sm)BaCuO composites. Mater. Sci. Eng. B (1999)
241. H.-P. Schiller, K. Grube, B. Gemeinder, H. Reiner, W. Schauer: Critical current homogeneity of Bi-2223 tapes determined by Hall-magnetometry. Inst. Phys. Conf. Ser. **158**, 981 (1997)
242. P. Schmitt, P. Kummeth, L. Schultz, G. Saemann-Ischenko: Two-dimensional behavior and critical current anisotropy in epitaxial $Bi_2Sr_2CaCu_2O_{8+x}$. Phys. Rev. Lett. **56**, 6237 (1997)
243. R. Schöffler, G. Pabst, J. Kellers: Progress in HTS wire and application development. Inst. Phys. Conf. Ser. **158**, 1267 (1997)

244. T. Schuster, H. Kuhn, A. Weisshardt, H. Kronmüller, B. Roas, O. Eibl, M. Leghissa, H.-W. Neumüller: Current capability of filaments depending on their position in $(Bi,Pb)_2Sr_2Ca_2Cu_3O_{10+x}$ -multifilament. Appl. Phys. Lett. **69**, 1954 (1996)
245. V. Selvamanickam: Talk presented at the ICMC/CEC 1997, Portland, OR (1997)
246. R. Semerad, B. Utz, P. Berberich, W. Prusseit, H. Kinder: Coevaporation of $YBa_2Cu_3O_{7-\delta}$ films up to 9 inches diameter. Inst. Phys. Conf. Ser. **148**, 847 (1995)
247. T. P. Sheahen: *Introduction to High-Temperature Superconductivity*. Plenum, New York, 1994
248. T. Shibata, J. Fujikami, S. Isjima, K. Sato, H. Ishii, T. Hara: Development of 3-phase 1 kA class high T_c superconducting power cable prototype. In *Advances in Superconductivity VIII* (eds. H. Hayakawa, Y. Enomoto), p. 1, Springer, Tokyo, 1996
249. J. Shimoyama, Y. Nakayama, K. Kitazawa, K. Kishio, Z. Hiroi, I. Chong, M. Takamo: Strong flux pinning up to liquid nitrogen temperature discovered in heavily Pb-doped and oxygen controlled Bi-2212 single-crystals. Physica C **281**, 69 (1997)
250. N. Siegel: Status of the large hadron collider and magnet program. IEEE Trans. Appl. Supercond. **7**, 255 (1997)
251. B. Skriba: *Abscheidung dicker YBCO-Filme*. Diplomarbeit, WU D 98-10, Bergische Universität GH-Wuppertal (1998)
252. D. Sommer, W. Kleemann, M. Lehndorff, K. Dransfeld: Microwave induced Brillouin scattering in $K_{1-x}Li_xTaO_3$. Solid State Comm. **72**, 731 (1989)
253. H. Suo, J.-Y. Genaud, G. Triscone, E. Walker, M. Schindl, A. Passerini, F. Cleton, M. Zhou, R. Flükiger: Preparation and characterization of {100}(001) cube textured Ag substrates for in-plane oriented HTS tapes. Supercond. Sci. Technol. **12** (1999)
254. A. Takagi, T. Yamazaki, T. Oka, Y. Yanagi, Y. Itoh, M. Yoshikawa, Y. Yamada, U. Mizutami: Preparation of melt-textuered $NdBa_2Cu_3O_y$ with $Nd_4Ba_2Cu_2O_{10}$ addition. Physica C **250**, 222 (195)
255. M. Tsuchimoto, N. Homma, K. Matsuura, M. Matsuda: Numerical evaluation of levitation force of a thin film high T_c superconductor. In *Advances in Superconductivity IX* (eds. S. Nakajima, M. Murakami), p. 1389, Springer, Tokyo, 1996
256. M. Tsuchimoto, K. Matsuura, N. Homma, M. Matsuda: Numerical evaluation of levitation force of a stack with high T_c superconducting thin films. Conf. Materials and Mechanisms of High-Temperature Superconducotrs (M2S-HTSC-V), Bejing, 1997 **D9**, 9 (1997)
257. M. Tsuchimoto, H. Takeuchi, T. Honma: Numerical analysis of levitation force on a high T_c superconductor for magnetic field configuration. Trans. IEEE Jpn. **114-D**, 741 (1994)
258. C. C. Tsuei, J. R. Kirtley: Determination of pairing symmetry using phase-sensitive tunneling in tricrystal cuprates. J. Low Temp. Phys. **107**, 445 (1997)
259. H. Tsuruga: State of development of the superconducting Maglev-system. In *Advvances in Superdonductivity VIII* (eds. H. Hayakawa, Y. Enomoto), p. 1273, Springer, Tokyo, 1996
260. M. Ulrich, D. Müller, K. Heinemann, L. Niel, H.-C. Freyhardt: Appl. Phys. Lett. **63**, 406 (1993)
261. P. Vase, et al.: High T_c update **12(10)** (1998)

262. M. V. Volkozub, A. D. Caplin, H. Eckelmann, M. Quilitz, R. Flükiger, W. Goldacker, G. Grasso, M. D. Johnston: Current distribution in multifilamentary HTS conductors. Inst. Phys. Conf. Ser. **158**, 1263 (1997)
263. P. Wagner, U. Frey, F. Hillmer, H. Adrian: Evidence for a vortex-liquid vortex-glass transition in epitaxial $Bi_2Sr_2Ca_2Cu_3O_{10}$ thin films. Phys. Rev. B **51**, 1206 (1995)
264. W. G. Wang, J. Horvat, J. N. Li, H. K. Liu, S. X. Dou: Effect of $Bi,Pb_3Sr_2Ca_2CuO_y$ phase on critical current density of $Ag/Bi,Pb_2Sr_2Ca_2Cu_3O_{10}$ tapes. Physica C **297**, 1 (1998)
265. W. G. Wang, J. Horvat, B. Zeimetz, H. K. Liu, S. X. Dou: Effect of $Bi,Pb_2Sr_2CuO_6$ phase on critical current density of $Ag/Bi,Pb_2Sr_2Ca_2Cu_3O_{10}$ tapes. Physica C **291**, 1 (1997)
266. Y. C. Wang, W. Biau, Y. Zhu, Z.-K. Cai, D. O. Welch, R. C. Sabatini, M. Suenaga, T. R. Thurston: A kinetic mechanism for the formation of aligned $Bi,Pb_2Sr_2Ca_2Cu_3O_{10}$ in a powder-in-tube processed tape. Appl. Phys. Lett. **69**, 580 (1996)
267. D. Wehler: *Herstellung von $YBa_2Cu_3O_{7-\delta}$ Schichten mittels Elektrophorese-Einfluss verschiedener Parameter auf den Oberflächenwiderstand*. Diplomarbeit, WUD 90-16, Bergische Universität GH-Wuppertal (1990)
268. R. Weinstein, Y. Ren, J. Liu, I. G. Chen, R. Sawh, V. Obot, C. Foster: Progress in j_c, pinning and grain size for trapped field magnets. In *Proc. Int. Symp. Supercond., Hiroshima, ISS93*, 1993
269. J. D. Whitler, R. S. Roth: *Phase Diagrams for High-T_c Superconductors*. American Ceramic Society, Westerville OH, 1991
270. R. Wilberg: *Kritische Stromdichte und U-I- Kennlinien von silberumhüllten BSCCO- Bandleitern und Drähten im Magnetfeld*. Diplomarbeit, WU D 96-36, Bergische Universität GH-Wuppertal (1996)
271. I. Wilke, G. Nakielski, J. Kötzler, K. O. Subke, C. Jaekel, F. Hüning, H.G.Roskos, H. Kurz: Dynamic conductivity of YBCO thin films from. Inst. Phys. Conf. Ser. **158**, 121 (1997)
272. M. K. Wu, J. R. Ashburn, C. J. Torng, P. H. Hor, R. L. Meng, L. Gao, Z. J. Huang, Y. Q. Wang, C. W. Chu: Superconductivity at 93 K in a new mixed phase Y–Ba–Cu–O compound system at ambient pressure. Phys. Rev. Lett. **58**, 908 (1987)
273. X. D. Wu, S. R. Foltyn, P. Arendt, J. Townsend, C. Adams, I. H. Campbell, R. Tiwani, Y. Coulter, D. E. Peterson: High current $YBa_2Cu_3O_{7-\delta}$ thick films on flexible nickel substrates with textured buffer layers. Appl. Phys. Lett. **65**, 1961 (1994)
274. W. Xing, B. Heinrich, H. Zhou, R. A. Crapp: Appl. Phys. Lett. **76**, 424 (1994)
275. Y. Yamada, B. Obst, R. Flükiger: Supercond. Sci. Technol. **4**, 165 (1991)
276. Y. Yamada, J. Q. Xu, E. Seibt, W. Goldacker, W. Jahn, R. Flükiger: Microstructural and transport properties of high j_c Bi2223/Ag tapes. Physica C **185** (1991)
277. C.-Y. Yang, S. E. Babcock, A. Goyal, M. Paranthamam, F. A. List, D. P. Norton, D. M. Kroeger, A. Ichinose: Microstructure of electron-beam-evaporated epitaxial yttria-stabilized zirconia/CeO_2 bilayers on biaxially textured Ni tape. Physica C **307**, 87 (1998)
278. C.-Y. Yang, A. Pashitski, A. Polyanskii, D. C. Larbalestier, S. E. Babcock, A. Goyal, F. A. Lee, C. Park, M. Paranthmam, D. P. Norton, D. F. Lee, D. M. Kroeger: Microstructural homogeneity and electromagnetic connectivity of $YBa_2Cu_3O_{7-x}$ grown on rolling-assisted biaxially textured coated conductor substrates. Physica C **329**, 114 (2000)

279. S. Yoo, M. Murakami, N. Sakai, T. Higuchi, S. Tanaka: Enhanced T_c and strong flux pinning in melt-processed $NdBa_2Cu_3O_x$ superconductors. Jpn. J. Appl. Phys. **33**, L1000 (1994)
280. B. Zeimitz, H. K. Liu, S. X. Dou: Microstructural study of Bi-2223/Ag tapes made using a two-stage sintering procedure. Supercond. Sci. Technol. **11**, 505 (1998)
281. M. Zeisberger, T. Klupsch, W. Michalke: Frequency, field, and temperature dependence of the AC penetration depth of a $GdBa_2Cu_3O_{7-\delta}$ film in the mixed state. Physica C **250**, 389 (1995)
282. E. Zeldov, D. M. M. Konczykowski, V. B. Geshkenbein, V. M. Vinokur, H. Shtrikman: Thermodynamic observation of first-order vortex lattice-melting transition in $Bi_2Sr_2CaCu_2O_8$. Nature **375**, 373 (1995)
283. W. Zhang, E. E. Hellstrom, M. Polak: A study of Ag-sheathed Bi-2212 tapes. Supercond. Sci. Technol. **11**, 971 (1996)
284. W. Zhang, M. Polak, A. Polyanskii, E. E. Hellstrom, D. C. Larbalestier: Formation of porosity and its effects on the uniformity of j_c in Bi-2212/Ag tapes. In *Advances in Cryogenic Engineering 44B* (eds. U. Balachandran, et al.), p. 509, Plenum, New York, 1998
285. W. Zhu, P. Nicholson: The effect of oxygen pressure on the formation of $Bi,Pb_2Sr_2Ca_2Cu_3O_{10+x}$. J. Mater. Res. **7**, 38 (1992)
286. W. Zhu, P. Nicholson: The influence of oxygen partial pressure and temperature on Bi–Pb–Sr–Ca–Cu–O 110 K superconductor phase formation and its stabiltity. J. Appl. Phys. **73**, 8423 (1993)
287. M. Ziese, P. Esquinazi, H. F. Braun: What do we learn from vibrating high temperature superconductors? Supercond. Sci. Technol. **7**, 869 (1994)

Index

14:24 alkaline-earth cuprate 31
14:24 phase 42
3-D positioning system 139, 167
3-D scanning setup 149
3221 42
3321 43

Abrikosov lattice 65
Abrikosov vortex 55, 57
ac field 117
ac susceptibility 112
activation energy 136
air bearing 159
alignment 46, 48, 93, 108
– biaxial 47, 117
alkaline-earth cuprate 42, 92
α 60
aluminum 47
amorphous layer 45
anisotropy 20, 51, 133, 134
– anisotropy parameter 20
application 5, 11, 117
Arrhenius behavior 135
atmosphere 46, 89
– oxygen 48
attractive force 148, 153, 164

BCS theory 52
Bean cone 63, 149, 169
Bean critical-state model 62, 146, 156
bearing parameter 158
$Bi_2Sr_2Ca_{n-1}Cu_nO_{4+2n-1}$ 31
$Bi_2Sr_2CaCu_2O_8$ 5, 17
Bi–Sr–Ca–Cu–O 17
Bi-2201 19, 45, 110, 115
Bi-2212 5, 19, 31, 85, 97, 110, 115
Bi-2212/Ag 101
Bi-2212/Ag conductor 88
Bi-2223 5, 19, 31, 85, 97, 110, 115
Bi-2223 formation 45
Bi-2223/Ag tape 7, 93, 172
Bi-3221 110, 115

biaxial alignment 47, 117
binary subsystem 26
Biot–Savart law 169, 174
brass texture 47
brick wall model 9
Brillouin zone 60
BSCCO conductor 85
bubble formation 85, 90
buffer layer 47, 48, 117, 122

Ca_2CuO_3 31
calcium plumbate 32
carbon 41
carbonate 41
CeO_2 47, 122
ceramic superconductor 7
cerium oxide 47
chemical barrier 48
chemical potential 23
coated conductors
– YBCO 46, 117
coherence length 7, 8, 54, 57
collective pinning 163
collective-creep model 67
component 22
composite material 118
conductor
– Bi-2212 101
– multifilamentary 86
connectivity 112
constantan 47
cooling
– intermediate 110
cooling rate 89
Cooper pair 53, 58
copper 47
– content 94
copper-free compound 92
copper-free phase 31, 106
core
– ceramic 92
correlation length

– pinning   68, 131
correlation volume   161
coupling strength   134
cracks   95
creep rate   65, 66
– effective   66
critical current   6, 97
– intergrain   7
– intragrain   7
– pair breaking   6
– percolation   10
critical current density   6, 47, 54, 62, 88, 102, 110, 129, 135, 156, 160, 163, 167, 169
– distribution   172
– engineering   104
critical field   54, 58
– lower   55, 57
– thermodynamic   55
– upper   55, 57
critical state   see Bean critical state
critical-current
– anisotropy   133
critical-current limitations   5
crystal growth   20, 46, 92
crystal structure   18, 20
– Bi–Sr–Ca–Cu–O   19
– YBCO   18
Cu–O chain   18
cube texture   47, 121
CuO_2 planes   18, 19
current   8
– intergrain   8
– intragrain   8
current leads   11
current path   10, 45, 120, 134
– percolative   10, 132
current transport   129, 173
current–voltage characteristics   130
current-carrying capacity   6

decoupling   135
decoupling transition   61
degree of freedom   24
demagnetization effect   63, 152, 156
density   85
– packing   91
deposition rate   122
diamagnetic model   142, 157
diamagnetic signal   113
DIC   121
differential interference contrast   121
dip coating   39

dipole field   152
directional solidification   36, 38
displacement   160
– irreversible   160
– reversible   160
drawing   41, 85
drift velocity   66
dwell time   89

E–j curve   131
E–j characteristics   67
EBSD   121
EDX
– line scan   118
EDX data   112
effective mass   51
elastic limit   160
elastic matrix   59
elastic properties   162
elastic regime   160
electron backscatter diffraction   121
electron microscopy   92, 102, 112, 123
electron–phonon interaction   53
energy
– elastic   59
– phase boundary   57
– self   60
energy gap   53
energy storage   13
energy technology   5, 12
entropy   23
epitaxy   48, 117
equilibrium   22
– phase equilibrium   22
– thermodynamic equilibrium   22
equilibrium reaction   33
eutectic   25
– eutectic point   25
– eutectic trough   26
– ternary eutectic   26
exponent
– dynamic   131
– static   131

fabrication
– conductor   38
– tape   38
– wire   38
fault current limiter   13
fcc metal   47, 117
Fermi surface   51
ferromagnetic   117
field-cooled   148, 153
filament   103

filling   41
filling density   41
film stack   164
fluctuation   60
flux bundle   65
flux creep   66
- thermally activated   65, 130
flux density   149, 154, 164
flux distribution   149, 167, 169, 173, 175
flux entry   105
flux flow   66
flux line   7, 55, 60, 148, 152, 159
- dynamics   62
flux line lattice   57, 59, 130, 133, 162
flux map   149, 174
flux motion   136
flux pattern   175
flux penetration   105
flux pinning   8, 62, 148
flux profile   63
flux quantum   57
fluxon   56
flywheel   13
force distribution   149
force sensor   139
- strain gauge   140
force–distance characteristics   147, 152, 163, 164
force–distance curve   153
force–distance hysteresis   158
forming gas   122
free enthalpy   23
freeway model   10

G–L equation   54
G–L theory   54
generator   12
geometry   140, 142, 146, 152
Gibbs free energy   23
Gibbs phase rule   24
Ginsburg number   60
Ginsburg–Landau equation   54
Ginsburg–Landau parameter   55, 57
Ginsburg–Landau theory   54
GLAG theory   54, 65
grain   134
- misoriented   120
size   119
grain boundary   6, 8, 10, 33, 45, 113, 121, 129, 161
- high-angle   121
- low-angle   8

grain contact   46
grain growth   120
grain size   161
green tape   39
- Bi-2212   40
- Bi-2223   40
green wire   39

Hall probe   167
Hall sensor   140, 149, 167
hastalloy   47
H_c see critical field
high-temperature superconductivity   5
HTSC   5
hysteresis   64, 145, 164
hysteresis losses   47, 117
hysteretic behavior   148, 164

I–V curve   130, 135
IBAD   47, 48
impurities   8, 59
- carbon   90
impurity phase   108
inclusions   8
incongruent melting   25, 32, 92
induction   152
inductive method   171
intercalation   46
interdiffusion   48
intermediate
- pressing   99
- rolling   99
intermediate state   57
internal energy   23
intrinsic mechanism   8
ion-beam-assisted deposition   47, 48
irradiation damage   7, 8
irreversibility field   8, 130, 136

j_c see critical current density
Josephson vortex   57
jump rate   66

κ 55
Kikuchi line   121
Kim–Anderson correction   172
Kim–Anderson model   65, 67, 135, 156

Labusch parameter   60, 161
λ 54
lateral resolution   149
lattice constant   117
lattice distortion   8, 19, 37, 59
lattice parameter   20, 48

Lawrence–Doniach model   56
lead   42, 45
lead content   8, 42, 46
lead solubility   42
lever rule   25
levitation force   139, 140, 144, 146, 147, 149, 152, 153, 158, 163, 164
Lindemann criterion   61
line tension   60, 135
liquid phase   95, 97, 112
liquid-phase processing   35
liquidus line   24
liquidus surface   26
London equation   54
London penetration depth   54
London theory   54, 146
loop
– hysteresis loop   153
– inner loop   153
– minor loop   153, 158
Lorentz equation   154
Lorentz force   59, 65, 129, 146, 152, 160
losses
– intergrain   114
LPP   35
LTSC   7

magic numbers   86
MAGLEV   13
magnet   11
– cryogen-free magnet   11
– dipole magnet   12
– high-field magnet   12
– quadrupole magnet   12
magnet diameter   142
magnet length   142
magnet technology   7, 11
magnetic angle   133
magnetic bearing   13, 139
magnetic field   5, 57, 129, 152, 174
– alternating   62, 160
magnetic flux   55, 63, 105, 139, 148, 153, 159
magnetic flux line   7
magnetic induction   63
magnetic levitation   13
magnetic stiffness   157, 159, 164
– vertical   158
magnetic suspension   139
magnetization   62, 143, 152, 171
– remanent magnetization   170
– reversible   62
magnetization curve   62, 152

magnetization model   153
magneto-optical imaging   105, 134
magnetotransport   129
master curve   68, 132
material processing   11, 17
Maxwell equations   146
Maxwell stress tensor   143
mechanical deformation   41, 85
Meissner state   58, 148
Meissner–Ochsenfeld effect   57, 139
melt powder melt growth   36
melt processing   34, 88
melt-grown $YBa_2Cu_3O_{7-\delta}$ 146, 158, 164
melt-textured growth   34
metal–insulator transition   52
metallurgy   117
microcracks   10
microcrystals   92
microscope
– optical   120
microstructure   20, 92, 102, 112
mirror charge method   146
misalignment   133
misalignment angle   133
misorientation angle   8
mixed crystal   20
mixed state   57
modulus
– compressional   60, 162
– shear   60, 162
– tilt   60
monocore wire   42
morphology   124
mosaic spread   49, 122
motor   12
MPMG   36
multifilamentary conductors   42
multifilaments   42
multilayer   48

Nb_3Sn 5
NbTi   5
Nd-123   38
Ni–Cu   47, 117
nickel   47, 117
– alloy   47, 117
– content   117
– substrates   122
nondestructive evaluation   173
nondestructive testing   167, 171
nucleation   46
nucleation center   38

order parameter
– glassy 68
orientation 121
orientational image mapping 121
orthorhombic 18
overpressure processing 101–103
oxidation 48
oxygen content 18, 33, 46
oxygen deficiency 8
oxygen vacancy 18, 19

PAIR process 39
pairing state 53
pancake vortex 56, 60, 135
particle accelerator 11, 12
$(Pb,Bi)_2Sr_2Ca_2Cu_3O_{10}$ 5, 17, 31
pellet
– sintered 146
penetration depth 54
– Campbell 60
– London 156
penetration field 154
percolation 10, 129
peritectic point 29
peritectic reaction 25, 29
peritectic temperature 35
permanent magnet 139, 152
persistent currents 153
phase 22
phase boundary 22
phase content 97
phase diagram 7, 22
– base 26
– Bi–Sr–Ca–Cu–O 30
– binary 24
– magnetic 58
– pseudo binary 29
– pseudobinary 27
– quaternary 30
– Y–B–C–O 28
phase formation 22, 42, 45
phase transition 60, 130, 133
– first-order 61
– second-order 61
pinning center 7, 37, 59, 65
pinning efficiency 149
pinning energy 160
pinning force 153
pinning potential 65
– harmonic 65, 160
– zigzag 65
pinning properties 7, 149, 159
pinning site 7, 49

PIT process 39, 85, 101
platelet 20, 92
PLD 47
plumbate 97
polarized light 120
pole figure 122
– inverse 122
porosity 101, 103, 108
potential
– parabolic 60
– pinning 60
power law 68
preannealing 101
precipitation 7, 37
precursor powder 39, 41, 85, 89
preparation 85
pressing 41, 86
– intermediate 95
pretreatment 41, 90
primary phase field 28
processing steps 96, 99
– final 109
profile rolling 85
properties
– normal-state 51
– physical 51
– superconducting 52
proximity effect 52
pseudoternary section 31
pulsed laser deposition 47

quality control 178
quartz balance 122
quaternary systems 28
quench study 89

RABiTS^{TM} 47, 49
radioactive irradiation 7
railway switch model 9
rate equation 66
RE–Ba–Cu–O 17
RE-123 38
reaction kinetics 22, 27, 46, 94, 97
reaction time 95
recrystallization 35, 38, 47, 89, 117
– Ni–Cu composite 119
– nickel 119
reduction ratio 47
refrigerator 11
relaxation time 68
remanent field 140, 144
repulsive force 148, 152, 164
residual phases 45
resistive barrier 73

resistivity 52, 67, 131
– flux creep 135
– flux flow 67
restoring force 61, 160
rolling 41, 47, 86
– intermediate 95, 109

scaling 68, 130, 131
scanning Hall probe 167
screening 113, 164
screening current 153, 157, 170
second phase 10, 37, 42, 106, 109, 112
seed crystal 38
SEM micrograph 102, 112
sheath material 39, 41
Shubnikov phase 7, 54, 55, 57, 62, 159
silver 41, 42, 47, 117
silver alloy 41
silver sheath 43, 45, 86
– alloying 72
– bridging 73
silver tape 173
silver tube 85
silver–BSCCO interface 88, 134
silver/Bi-2212 interface 108
single-vortex pinning 60, 162
sintered sample 158
sintering 109
Sm-123 38
solid phase 24
solid-state diffusion 37
solid-state reaction 93
spatial resolution 167
spin correlation 52
spin fluctuation 52
spring constant 160
sputter technique 117
– high-frequency magnetron 122
$Sr_{14-x}Ca_xCu_{24}O_{40-y}$ 31
stiffness 158, 163
stoichiometric composition 28
stoichiometry 41, 89
– cation 41, 94
strong link 8
substrate heater 122
superconducting levitation 150
superconductor
– anisotropic 54
– hard 7, 59, 62, 136, 139, 148
– layered 56
– low-temperature 7
– type I 62, 139
– type II 7, 57, 62, 139, 148

supercurrent 6
surface current 153
surface impedance measurements 53
susceptibility
– third-harmonic 163
suspension 153
symmetry 53
– d-wave 53
– order parameter 53
– p-wave 53

tape
– Bi-2212/Ag 103
– Bi-2223 133
– Bi-2223/Ag 109
– long tape 178
– monocore 85
– multifilamentary 86
– Ni–Cu composite 124
– recrystallized 119
temperature gradient 36
temperature–time diagram 35, 89, 101
test magnet 142
tetragonal 18, 20
texture 88, 92, 108, 122, 134
– analysis 120
– cube texture 117
– in-plane 49
– out-of-plane 49
texture formation 46
thermal processing 88
thermal treatment 39
thermo-mechanical processing 46, 85, 93
thermo-mechanical treatment 41, 42, 95
thermodynamic potential 23
thermodynamics 22
thick film 47, 49
thickness of tape 88, 104
thin film 158
– epitaxial 146, 163
– stack 163
tie line 26
top-seeding technique 38
transformer 12
transient liquid 45
transmission cable 12
TRANSRAPID 13
trapped field 168
trapped flux 150, 169
tricrystal 53
tunneling spectroscopy 53
twin formation 19

two-band model   53

uninterrupted power supply   13
UPS   13

virgin curve   152, 153
viscous motion   61
void   10, 89, 101, 103
vortex   7, 129, 153
– pancake vortex   7
vortex core   59
vortex dynamics   65
vortex glass   68, 130
vortex glass model   67, 135
vortex lattice   60
vortex liquid   61, 68, 130
vortex motion   65, 129
vortex solid   61
vortex spacing   60, 162
vortex–vortex interaction   162

weak link   8, 129

wire
– Bi-2212   136
– multifilamentary   85

x-ray diffraction analysis   33, 94, 97, 107, 110, 115
ξ 7, 54

Y-123   18, 28
– growth of Y-123   37
Y-211   20, 28
– inclusions   20, 37
– Y_2BaCuO_5 8
$YBa_2Cu_3O_{7-\delta}$ 8, 17, 34
$YBa_2Cu_3O_{7-\delta}$ ceramics   33
YBCO   18
YBCO film   49, 117
– epitaxial   171
YSZ   47, 123

zero-field-cooled   147
zirconia   47

Springer Tracts in Modern Physics

150 **QCD at HERA**
The Hadronic Final State in Deep Inelastic Scattering
By M. Kuhlen 1999. 99 figs. X, 172 pages

151 **Atomic Simulation of Electrooptic and Magnetooptic Oxide Materials**
By H. Donnerberg 1999. 45 figs. VIII, 205 pages

152 **Thermocapillary Convection in Models of Crystal Growth**
By H. Kuhlmann 1999. 101 figs. XVIII, 224 pages

153 **Neutral Kaons**
By R. Belušević 1999. 67 figs. XII, 183 pages

154 **Applied RHEED**
Reflection High-Energy Electron Diffraction During Crystal Growth
By W. Braun 1999. 150 figs. IX, 222 pages

155 **High-Temperature-Superconductor Thin Films at Microwave Frequencies**
By M. Hein 1999. 134 figs. XIV, 395 pages

156 **Growth Processes and Surface Phase Equilibria in Molecular Beam Epitaxy**
By N.N. Ledentsov 1999. 17 figs. VIII, 84 pages

157 **Deposition of Diamond-Like Superhard Materials**
By W. Kulisch 1999. 60 figs. X, 191 pages

158 **Nonlinear Optics of Random Media**
Fractal Composites and Metal-Dielectric Films
By V.M. Shalaev 2000. 51 figs. XII, 158 pages

159 **Magnetic Dichroism in Core-Level Photoemission**
By K. Starke 2000. 64 figs. X, 136 pages

160 **Physics with Tau Leptons**
By A. Stahl 2000. 236 figs. VIII, 315 pages

161 **Semiclassical Theory of Mesoscopic Quantum Systems**
By K. Richter 2000. 50 figs. IX, 221 pages

162 **Electroweak Precision Tests at LEP**
By W. Hollik and G. Duckeck 2000. 60 figs. VIII, 161 pages

163 **Symmetries in Intermediate and High Energy Physics**
Ed. by A. Faessler, T.S. Kosmas, and G.K. Leontaris 2000. 96 figs. XVI, 316 pages

164 **Pattern Formation in Granular Materials**
By G.H. Ristow 2000. 83 figs. XIII, 161 pages

165 **Path Integral Quantization and Stochastic Quantization**
By M. Masujima 2000. 0 figs. XII, 282 pages

166 **Probing the Quantum Vacuum**
Pertubative Effective Action Approach in Quantum Electrodynamics and its Application
By W. Dittrich and H. Gies 2000. 16 figs. XI, 241 pages

167 **Photoelectric Properties and Applications of Low-Mobility Semiconductors**
By R. Könenkamp 2000. 57 figs. VIII, 100 pages

168 **Deep Inelastic Positron-Proton Scattering in the High-Momentum-Transfer Regime of HERA**
By U.F. Katz 2000. 96 figs. VIII, 237 pages

169 **Semiconductor Cavity Quantum Electrodynamics**
By Y. Yamamoto, T. Tassone, H. Cao 2000. 67 figs. VIII, 154 pages

170 **d-d Excitations in Transition-Metal Oxides**
A Spin-Polarized Electron Energy-Loss Spectroscopy (SPEELS) Study
By B. Fromme 2001. 53 figs. XII, 143 pages

171 **High-T_c Superconductors for Magnet and Energy Technology**
By B. R. Lehndorff 2001. 139 figs. XII, 209 pages

Printing: Mercedes-Druck, Berlin
Binding: Stürtz AG, Würzburg